T0074491

Exotic Amphibians and Reptiles of the United States

UNIVERSITY PRESS OF FLORIDA

Florida A&M University, Tallahassee
Florida Atlantic University, Boca Raton
Florida Gulf Coast University, Ft. Myers
Florida International University, Miami
Florida State University, Tallahassee
New College of Florida, Sarasota
University of Central Florida, Orlando
University of Florida, Gainesville
University of North Florida, Jacksonville
University of South Florida, Tampa
University of West Florida, Pensacola

EXOTIC AMPHIBIANS AND REPTILES OF THE UNITED STATES

WALTER E. MESHAKA JR., SUZANNE L. COLLINS,
R. BRUCE BURY, AND MALCOLM L. MCCALLUM

University Press of Florida

Gainesville · Tallahassee · Tampa · Boca Raton

Pensacola · Orlando · Miami · Jacksonville · Ft. Myers · Sarasota

27 26 25 24 23 22   6 5 4 3 2 1

Library of Congress Cataloging-in-Publication Data
Names: Meshaka, Walter E., 1963- author. | Collins, Suzanne L., author. | Bury, R. Bruce, author. | McCallum, Malcolm L., author.
Title: Exotic amphibians and reptiles of the United States / Walter E. Meshaka, Suzanne L. Collins, R. Bruce Bury, Malcolm L. McCallum.
Description: Gainesville : University Press of Florida, 2022. | Includes bibliographical references and index. | Summary: "The first complete field guide to the exotic amphibians and reptiles established in the continental United States and Hawaii, this book provides practical identification skills and an awareness of the environmental impacts of these species"— Provided by publisher.
Identifiers: LCCN 2021040743 (print) | LCCN 2021040744 (ebook) | ISBN 9780813066967 (hardback) | ISBN 9780813057859 (pdf)
Subjects: LCSH: Amphibians—United States. | Reptiles—United States. | Exotic animals—United States. | BISAC: SCIENCE / Life Sciences / Zoology / Ichthyology & Herpetology | NATURE / Animals / Reptiles & Amphibians
Classification: LCC QL652 .M42 2022 (print) | LCC QL652 (ebook) | DDC 597.80973—dc23
LC record available at https://lccn.loc.gov/2021040743
LC ebook record available at https://lccn.loc.gov/2021040744

The University Press of Florida is the scholarly publishing agency for the State University System of Florida, comprising Florida A&M University, Florida Atlantic University, Florida Gulf Coast University, Florida International University, Florida State University, New College of Florida, University of Central Florida, University of Florida, University of North Florida, University of South Florida, and University of West Florida.

University Press of Florida
2046 NE Waldo Road
Suite 2100
Gainesville, FL 32609
http://upress.ufl.edu

WEM: To Betty S. Ferster, my wife and one and only, who's made so much of everything worth it.

SLC: To Joseph T. Collins (1939–2012), whose love for amphibians and reptiles lives on through the many students he inspired.

RBB: To the next generation of biologists, who not only need to study these species but also to check them.

MLM: To the many victims of the COVID-19 pandemic. May the losses remind us that ignorance cannot shield one from reality.

Contents

PART 5. SNAKES 203

Burmese Python Colonization and Eradication: A Stitch in Time Unnecessarily Missed in the Everglades 205
Michael R. Rochford

PART 6. CROCODILIANS 223

Acknowledgments

We are grateful for the professional kindnesses of Susan Barnes of the Oregon Department of Fish and Wildlife, Craig Guyer of Auburn University, Marc Hayes of Washington Department of Fish and Wildlife, Wil Pitt of the Smithsonian National Zoological Park, and Robert Powell of Avila College. For photographs of the species, we thank Ronald Altig, Torsten Blanck, Gary Busch, Tien-His Chen, Ryan Choi, Dick Culbert, Bernard Dupont, Kevin M. Enge, William L. Farr, Paul Freed, Brian Gratwicke, Shi Haitao, K. M. Haneesh, Daniel F. Hughes, Dennis Jarvis, Janson Jones, Raju Kasambe, Fred Kraus, Jonathan Losos, Melissa Losos, Bill Love, Er. Ameet Mandavia, Vicente Mata-Silva, Ian R. McCann, Gerald McCormack, Colin Meurk, Omid Mozaffari, Carlos Nieves, Michael Rochford, Jake Scott, Ianaré Sévi, Steve Taylor, Simon Tonge, Jacinta Lluch Valero, Eugene Wingert, and Waifer X. We extend our gratitude to the staff of the University Press of Florida for their guidance and assistance throughout this project.

Introduction

In this book, we provide information on the identification, history of introduction, geographic ranges, and ecology of exotic species of amphibians and non-avian reptiles established in the continental United States and Hawai'i. Our goal is to provide an up to date documentation of the exotic herpetofauna as a tool for management professionals, researchers, citizen scientists, and naturalists throughout the country who need or want to track, understand, and manage this segment of the biota.

We undertook this endeavor because the topic of exotic species is important, and documentation of the established species is timely. Exotic species constitute a worldwide phenomenon, one that is evident in the many exotic plants and animals found in residential backyards. What constitutes an exotic species? Very simply, an exotic species is one whose dispersal is a result of human mediation. This is when humans, rather than natural dispersal, are responsible for the presence of the species in a given locality. Norway Rat, *Rattus norvegicus* (unintentional), and feral hog, *Sus scrofa* (intentional), having arrived from Europe to the United States in Spanish galleons among other ships, are exotic. The Cattle Egret, *Bubulcus ibis*, having arrived in the United States from Africa by natural processes, is native but not endemic. The number of exotic plants and animals established in the United States is staggering. The U.S. Geological Survey recognizes more than 6,500 exotic species in the United States. The ecological and financial costs associated with this phenomenon can be severe for impacted systems, the economy, and health and human safety. Some regions of the United States support a battery of exotic species to the extent that segments of the exotic biota rival those of indigenous counterparts.

To better explain presence and impacts of these species, we offer definitions of included groups:

Exotic species—A non-native species whose presence is the result of human-mediated dispersal outside its indigenous geographic range.

Invasive species—Describes a species that causes ecological or economic harm in a new environment where it is not native.
Native species—A species whose presence is not the result of human mediation.
Endemic species—A species whose native range is restricted to one geographic location. Often employed to indicate a species with a small geographic range.

The second reason we undertook this project is a matter of timeliness. A perusal of Peterson field guides to reptiles and amphibians of the eastern United States—the third edition of Conant and Collins (1991) and that of Powell and colleagues (2016)—shows a marked upswing in the exotic herpetofauna established in the United States over a period of 25 years. Meshaka and colleagues (2004) reported 40 such species established just in Florida. Seven years later, Meshaka's (2011) list increased to 47 established species. Also, in Florida, Krysko and colleagues (2011) reported 137 exotic herpetofaunal taxa, 56 of which they considered established. McKeown (1996) documented an exotic herpetofauna consisting of 27 species in Hawai'i. With a broader geographic scope, Lever (2003) documented 272 exotic herpetofaunal species worldwide, and Kraus (2009) provided introduction records for 676 such taxa worldwide. Each contribution remains of value; however, all are out of date. Moreover, criteria used to determine establishment of a species can vary among authorities. Consequently, the importance of the issue and need for a synthesis and up-to-date, practical source that documents the presence and ecology of the exotic herpetofauna of the United States prompted the writing of this book.

Scope of the Issue

What taxonomic groups and how many species of each group are established in the United States?

The taxonomic representation of these exotic species is highly uneven (table 1). Among the 103 established exotic species, 74 species are exotic to the United States and 29 species are exotic to regions of the United States where they are elsewhere native. Lizards, with 62 species in 15 families, account for 60.2% of all exotic species of amphibians and reptiles of the United States. Two of the lizard species are native to other parts of the country. Geckos make up approximately one-third of the exotic lizard species, with 21 species. Most of these are of small to medium body sizes, generally insectivorous,

Table 1. Taxonomic representation of non-native and extralimital species of amphibians and reptiles identified as established exotic herpetofauna of the United States

Taxonomic group	Exotic non-native species	Extralimital native species	Total
Salamanders	0	3	3 (2.9%)
Frogs and Toads	7	11	18 (17.4%)
Turtles	2	10	12 (11.7%)
Crocodilians	1	0	1 (1.0%)
Lizards	60	2	62 (60.2%)
Snakes	4	3	7 (6.8%)
Total	74	29	103 (100.0%)

mostly nocturnal, and generally edificarian, residing in buildings. A few, however, are diminutive (e.g., the Ashy Gecko, which is ca. 2.75 in. long) whereas others are large (e.g., the Tokay Gecko, which reaches lengths of 12.0 in.). Among the geckos of the genus *Hemidactylus*, a taxon cycle is in play (in which species are replaced by others), and coexistence among these ecologically analogous species is highly unstable. Second in species richness to geckos are anoles (genus *Anolis*), which comprise 10 exotic species. Interactions among the anoles are complex, and competition, predation, and hybridization have all been documented. Skinks, with seven species, and teiids with six species, are the next most frequently represented exotic lizards. The remaining species are grouped into eight different families, some of which include very large representatives. Two Spiny-tailed Iguana species, Mexican Spiny-tailed Iguana and Gray's American Spiny-tailed Iguana, overlap ecologically to some extent with the Argentine Giant Tegu, Gold Tegu, the Green Iguana (a highly fecund, fast-maturing herbivore that can be abundant around water), and the Nile Monitor (ecologically somewhat analogous to the Argentine Giant Tegu and a large carnivore).

Frogs represent the second most speciose taxonomic group with 18 species, or 17.4% of the total number of species. This group is represented by six families. Eleven of the frog and toad species are native to other regions of the country. The true frogs (Ranidae), with nine species, are the best-represented anuran group. Within this group, eight of the species are native.

Turtles represent the third most speciose taxonomic group, with 12 species, or 11.7% of the total number. Ten of the turtle species are native to other parts of the country. Box and water turtles make up the majority of the turtles and are represented by six species, all native. Because of their popularity in the pet trade (mostly earlier) and continued illegal collecting, many species are caught and later released all over the United States. Most are single or a few individuals, but there is a high potential for establishment of breeding populations.

Snakes are represented by seven species, or 6.8% of the total number of species. Three of the snake species are native. Yet, the species within this small group range from the easily overlooked Brahminy Blindsnake to three nearly impossible to ignore large constrictors, the Boa Constrictor, Burmese Python, and Northern African Python.

Salamanders are represented by three native species: the Western Tiger Salamander, Seal Salamander, and Southern Two-lined Salamander, or 2.9% of the total number of species. Crocodilians are represented by one species, the Spectacled Caiman, an exotic Neotropical predator, that represents 1.0% of the total.

What are the native centers of distribution of the exotic herpetofauna?

Slightly more than half (60 of the 103 species) originate from the Western Hemisphere, with the other 43 native to the Eastern Hemisphere. Most of the species are native to the United States (29 species) and Asia (25 species). Geckos, with 10 species, represent the majority of Asian species. Eleven species of frogs and 10 species of turtles form the majority of native species that are exotic elsewhere in the United States. West Indian species comprise 17 species, 14 of which are lizards. Central and South American species combined comprise 14 species, 12 of which are reptiles. Africa, with 11 species, is represented by nine species of lizards. Madagascar, with four species, and Europe, with three species, are represented exclusively by lizards.

Which species are most successful?

Geographic distribution, as a measure of success, is greatest in the Mediterranean Gecko. It is a superb human commensal and found in 21 states, even as it retreats from some historic areas, where it is being replaced by more aggressive congeners, such as the Tropical House Gecko, Indo-Pacific House Gecko, Rough-tailed Gecko, and perhaps the Sri Lankan Spotted House Gecko. The Red-eared Slider, an inexpensive, common, and popular pet, is found in 20 states outside its native range in the central United States. The North American Bullfrog is found in 15 states outside its native range in the eastern United States. Adult frogs are prized for consumption, and tadpoles are used both as bait and for algal control in ornamental fishponds.

Second to geckos in number of established species are the anoles. This is an ecologically diverse group whose structural niches are easily recognized by associated morphological adaptations. With the exception of the Knight

Anole, an omnivorous large branch and crown giant, and the Jamaican Giant Anole, another large species, most anoles are much smaller in body size. The most successful of the anoles is the Brown Anole, which is established in eight states. Populations in the United States comprise a genetic melting pot from multiple native sources. It is a vigorous, broadly adaptable species firmly established in southeastern states and in Hawai'i. Many other species listed in this book are found in only one or two states and often in only a few counties. Occasionally, the few counties are far from one another owing to the human-mediated agency of animal dispersal. The same may be said for localized populations in disparate states, as in the cases of the Brahminy Blindsnake (nine states), the Italian Wall Lizard (eight states), and the African Clawed Frog (four states). The vast majority, however, are found in aggregates of counties or islands.

Which locations have the highest number of species?

The greatest number of exotic herpetofaunal species is in Florida (60 species) and Hawai'i (30 species). The two states share 15 species. For both states, a combination of commercial ports, amenable climate, and extensively disturbed habitat provide opportunities for exotics to arrive and become established. Furthermore, Hawai'i has no native terrestrial amphibians and reptiles that elsewhere occupy niches that could reduce the likelihood of successful colonization by exotics. Hawai'i's exotic herpetofauna is derived in part from incidental dispersal by the first Polynesians to settle the islands, intentional introductions of species for human consumption or agricultural pest control, and through the pet trade. Florida, with past lax regulations concerning exotic species, is at greater risk of one ecological atrocity after another and now supports an exotic herpetofauna that is overwhelmingly, even if not exclusively, pet trade–related in its derivation.

What are some of the ecological relationships between exotic and native species?

Not surprisingly, combining exotic species from around the world and placing them in environments with indigenous ecological communities can result in dramatically altered food webs. All the exotic species consume some to many native plants and animals. The Cuban Treefrog, for example, eats native invertebrates, native and exotic frogs, and exotic anoles and geckos. In turn, it is eaten by native snakes, like the Peninsula Ribbonsnake, *Thamnophis sauritus sackenii,* and the Everglades Ratsnake, *Pantherophis alleghaniensis rossalleni,* by native birds, such as the Little Blue Heron, *Egretta caerulea,* and Barred Owl, *Strix varia,* and by the exotic Knight Anole with which it naturally co-occurs in Cuba. The Burmese Python eats a wide range of native mammals and birds, negatively impacting populations of many species. Its trophic relationship with the American Alligator, *Alligator mississippiensis*, the top native predator of the Everglades, can be as predator or as prey.

How are exotic species moved around?

Human-mediated dispersal of these species can be either intentional or adventitious. Sometimes, dispersal can be of the same species to different places or at different times. Knight Anoles were intentionally released on the campus of a university in Miami, Florida. In time, individuals dispersed further by other, sometimes natural, means, although most dispersal occurs on vegetation transported by human agency. Other individuals are dispersed by humans who intentionally establish them on their properties, and still others escape from pet dealerships that carry them in counties they otherwise would not colonize on their own for a long time, if at all. As a matter of scale, the same principle applies among states and can be illustrated by the case of the Mediterranean Gecko. It was incidentally carried to Key West from Cuba on cargo, then individuals were incidentally carried within mainland Florida and other states in automobiles. Intentional releases to establish a colony to harvest for profit or as a source of residential pest control, in addition to accidental releases of individuals acquired as pets or as food for captive snakes, further increase opportunities for colonization. Intentional introductions for human consumption, as in the case of the North American Bullfrog and Wattle-necked Softshell, or agricultural pest control, such as the Cane Toad and Northern Curly-tailed Lizard, similarly contribute to further dispersal, even if for reasons that are different from those involved in initial introductions.

Natural processes can serve as dispersal agents for exotic species. Natural floods and hurricanes can move individuals farther and faster than they could otherwise go themselves. Hurricanes can also serve as a dispersal agent through the destruction of captive breeding facilities; hurricane-mediated dispersal can affect both large-scale outdoor breeding facilities as well as small indoor private breeding operations that can maintain a large breeding stock.

Criteria

How do we determine that a species is established?

Our criteria for an established species follow those of earlier works by one of us (Meshaka). First, a voucher must exist. Preferably, the voucher (= proof) is a preserved specimen or series of specimens documenting the location of a colony. Far less valuable is a photographic voucher. Either, however, must be verified by a trained herpetologist and deposited in a museum. Second, mixed size- and age-classes must be present. Third, evidence must exist that a colony's presence exceeds one presumed generation, which can be determined by subsequent visits or by credible testimony of local individuals. The absence of one or more of these three criteria does not negate the reality of establishment but is generally considered insufficient proof. For example, finding a few adult males and several gravid females at a site is worthy of concern; however, absence of mixed-size classes or no evidence of successful reproduction leaves open the possibility that this group of animals has thus far only survived with *potential* for establishment. Consequently, we exclude the Ocellated Skink, *Chalcides ocellatus*, because the age of the colony was unrecorded. Lacking a voucher from the only suspected colony, the Javanese File Snake, *Acrochordus javanicus*, is omitted from our list of established species. Further, many other species with only one or a few documented individuals could in actuality be established species. Such waifs can be diverse. For example, more than a dozen species of native and some exotic species of turtles are found in urban waterways from Seattle, Washington, south to San Diego, California. We excluded them because of lack of evidence of breeding. In most cases, no one has documented their status (i.e., breeding or not). It remains the work of researchers to clarify the status of these and other unconfirmed species.

How do we determine the geographic distribution of an exotic species?

Two choices exist in determining where these species, or any species for that matter, occur. One method is to rely exclusively on published records that reference a vouchered specimen. Another method is to use locations noted in the list of holdings of species stored as museum specimens. Ideally, the researcher should examine them to verify the accuracy of their identifications. Records such as these are sometimes taken at face value, except for those considered outliers from an expected range. With an interest in maintaining a permanent written record backed by vouchers that may or may not survive, we opted for the former approach, considering all published records through

2019. We strongly encourage publication of unknown records upon examination of specimens in museums. Noting the catalogue number eliminates duplicative publication of the same record. *Herpetological Review*, published by the Society for the Study of Amphibians and Reptiles, is replete with examples of this practice having been employed to good effect for distributional records and to document the status of both native and exotic species of amphibians and reptiles, often for a particular state. In turn, these written records are cited in publications or used in literature searches, as with this book. For county, parish, independent city, borough, and island-level accuracy, our literature searches ended at the end of the 2019 calendar year. In our references, we provide general, regional, and state-specific sources used in searches for locality records and other information associated with the species accounts. Island are listed and mapped by county.

Voucher specimens are analogous to a research library, each one a first-edition book, providing information capable of answering many pertinent questions and as a reservoir for questions unasked or even unknown without the discovery of new techniques, such as in genetics, disease, and toxicology. A series of specimens captures variability and provides even more information. Photographic vouchers are better than nothing and are the only recourse for documentation of a population that could not withstand the loss of an individual. A photograph may also be the only voucher possible if the individual of an important record is out of reach or collection of a specimen is illegal. Still, photographic vouchers are generally of limited value. As a matter of practice, we advocate the collection and deposition of specimens.

The nature of science is such that some things are known, others are yet to be determined, and all are subject to change. Thus, a voucher may or may not indicate establishment of a species but does reflect presence. To that end, we may be able to say that a certain species is established in Mobile, Alabama, for instance, but not know whether some or even all the other locations for which we have records represent reproductive colonies or merely presence of unknown status. Therefore, our approach is to include isolated records of waifs only in states for which at least one colony is verified. Thus, for example, an isolated county record of the Brown Anole derived from a nursery shipment to a big box store in a state not known to be colonized by Brown Anoles is not included in our maps. In this way, the reader is assured that a state with county records is one in which the species is established. Such situations are discussed in the following species accounts. We also provide a list of useful sources that can help you learn about a particular species. For the sake of conformity, we follow the most up to date taxonomic arrangements, with the exception of the Everglades Ratsnake, which we retain in recognition of its regional distinction. With four exceptions, our scientific and common names follow those of the 8th edition of *Scientific and Standard English Names of Amphibians and Reptiles of North America North of Mexico* (Crother 2017). For those exceptions, we use the common names in the 2016 *Peterson Field Guide to Amphibians and Reptiles of Eastern and Central North America* (Powell et al. 2016). One is for *Anolis cybotes,* alongside which we provide the SSAR name. The Green Frog, Bullfrog, and Green Anole are common names that apply to species found outside the United States, and this book represents an international herpetofaunal community. Consequently, we have elected to precede the common names of these three species with the appellation of North American. For easier and more practical discernment between the Snapping Turtle and the Alligator Snapping Turtle, *Macrochelys temminckii,* which are often called snapping turtles, we have chosen to refer to the Snapping Turtle as the Common Snapping Turtle.

Can museum vouchers be cited in a work without personal examination by the author(s) of that work?

The answer to this question depends on the level of accuracy associated with the publication. Data downloads and mapping of museum holdings are relatively easy in the digital age. Not surprisingly, the chance and extent of errors increase without examination of each specimen. Authors adopting this approach often verify outlying records. This is useful for field guides with the goal of providing a general map that is accurate to a regional level. It provides a starting point in understanding biogeography and at a basic level provides the reader with information needed to determine whether a suspected find is well outside a species' geographic range.

On the other hand, publications aiming to serve as a county-level source of accuracy fall short of the goal by not verifying the identity of each record. At best, such a publication could provide symbols to designate verified and unverified records. Ideally, catalogue numbers should be provided in an appendix. If this is not feasible, a range of collection dates from each institution can be helpful.

What is a report?

Whereas a record constitutes a vouchered specimen, a report is an observation, with the observer identified in the account. If represented by a different symbol from that of a record, reports can be useful, such as those found in publications or websites relating to a site's herpetofauna, and can serve as a starting point for verification with vouchers and field studies.

Conservation of Native Biota

What is to be done about exotic species?

With respect to exotic species, all the species in this book have in common some combination of ecological characteristics associated with a high likelihood of colonization success. These patterns have been corroborated time and time again: high fecundity, high vagility, short generation times, ability to function in a wide range of physical conditions, similarity between source and target areas, coexistence with humans, broad diets, open niche space/ few competitors, predator-free space/low predation pressure.

The approach of studying the life history of an organism within the context of reasons for its success or lack of success is the key to knowing how to avoid potentially successful species and can be helpful for knowing potential impacts and the extent to which they can expand their introduced range and be controlled, or if they can be eradicated. This has been done with some species for control, and to a lesser extent for prevention. Here, we proffer that value is untapped in using what is extremely well known—*collectively* for prevention and as *species-specific* for management.

As a preventive measure, several ecological correlates should be used as filters or criteria to reduce the risk of colonization. In this way, federal and state authorities, along with conservation organizations, would have an organized and quantifiable process to identify potentially successful exotic species and thereby determine which species would be safe to keep or to farm. To that end, it puts the onus on the advocate of an importation to demonstrate a low to nonexistent risk of colonization success. THOSE are solid moves to reaching a leveling off in the exotic species accumulation curve.

For the present:

1. Acquisition, protection, and maintenance of natural communities will pay dividends. Altered systems are more at risk of exotic species colonization than those that are mostly intact. This step can be achieved through protection of natural areas by public and private institutions.
2. Restrict sale of native amphibians and reptiles. The loss of these species breaks the integrity of the system. A dearth of predators and competitors can provide an advantage to colonization. This action step is within the purview of state agencies that regulate the take of game and nongame species.

3. Teach state history and ecology in grade and high school. AWARE-NESS will foster an ecological conscience and a conservation ethic. Both public and private schools can incorporate these lessons and can use local parks and historical sites for active learning and for meeting age-appropriate standards in biology and history.

4. Fund thorough life-history studies. Life-history data can be applied to anticipate with considerable accuracy the potential impacts, rate, and direction of dispersal by exotics. Such studies can can help identify methods for biocontrol as well as provide information for evaluating the likelihood of eradication. Federal and state grants focusing on this topic can be made a priority. The results of these studies should be shared in such venues as peer-reviewed journals and presentations to the public at large.

Humans should be responsible stewards of our natural legacy, one that provides clean air and water, one that provides immense joy to the heart and mind, and one upon which we can build our lives. If done responsibly, by living off the interest and not the capital of healthy systems, this is a measure of the best of humanity.

Organization of the Species Accounts

One hundred and three species accounts were assembled for 74 species that are exotic to the United States and for 29 native species that are exotic extra-limitally within the United States. Each account begins with a description of the species in which we provide an adult body length, with the description (small, medium-sized, large) relative to members of its taxonomic family. For salamanders, snakes, and crocodilians, body size is expressed in total length, or length from tip of the snout to the end of the tail. Body size of frogs and toads is measured in snout-vent length, which is the distance from the tip of the snout to the cloaca. Body size of turtles is measured in carapace length, which is the straight-line distance from the front to the rear of the carapace, or portion of the shell that is above the turtle. We use the English system of measurement. Sexual dimorphism in body size is noted. The taxonomic family of each species is provided, and we list the recognized subspecies of polytypic species that are native to the United States. Color and pattern as well as any differences in these traits among males, females, and juveniles are noted, and one or more representative photographs accompany the description. The geographic range, or more broadly speaking, the center of distribution is noted.

The next section discusses the history of introduction and the introduced geographic range. Where possible, we note the year a species was first introduced to, or detected in, a state where it colonized. Our criteria to assess establishment are presented in the Introduction. Because the first record of a species may be published years after it was first detected, we note both the year of the publication and the year of specimen collection. Geographic distribution maps were compiled from an examination of 2,908 county, parish, island, independent city, and borough records ending in December 2019. Sources for these records are listed in the references. Thus, a researcher wishing to publish new records can cite this book and any additional records published after 2019. A researcher interested in conducting a full review of a given species covered in this book can use the sources in the references as a starting point.

The third section in each account is ecology. There, we provide information on habitat, activity, reproduction, diet, predators, and potential and documented impacts. The reader will see a great deal of variability in this section, both within and among species, which reflects an overall scarcity of life history information in the target areas where species have been introduced. Information in this section can be predictive and testable by researchers interested in colonization dynamics and make a researcher aware of topics especially in need of investigation.

Interspersed among the major taxonomic headings we present eight essays that discuss aspects of the colonization phenomenon among exotic herpetofauna. It is hoped these essays are thought-provoking and informative and can provide insights on how humans affect exotic species, and how exotic species affect us. These essays, like this book itself, are progress reports to better guide humans in making wise decisions as stewards of the natural world.

PART 1

SALAMANDERS

Preceding the following salamander species accounts, we present an invited essay. In it, Malcolm McCallum discusses intended and unintended consequences of aquaculture of native and exotic species.

The Role of Aquaculture in the Problem of Exotic Species

Malcolm L. McCallum

Exotic and invasive species are important issues for the conservation and management of our natural resources. When an exotic is introduced into a new area, the interaction of its establishment with the fine network of interlacing relationships among the residing species is often impossible to predict. It is further even more difficult to project how these introduced organisms will interact with the geological, hydrological, chemical, and other physical processes that have developed in those ecosystems over the millennia. This makes it very important to recognize the value of biosecurity in captive rearing programs, regardless of their purpose.

Aquaculture is one form of animal production that is widely applied for commercial and conservation needs. From an aquatic perspective, the exotic organisms most widely known are game fish. For decades, state and federal aquaculture facilities produced game fish for introduction into surface waters in part to generate income from recreational fishing. These introductions created many problems for native fishes, particularly in the western watersheds where niches were largely occupied by different life stages of single organisms. Less is known about these impacts on herpetofauna. Certainly, the introduction of game fish into isolated small fishless ponds has taken its toll. Species that regularly breed in fish-occupied ecosystems typically have some defense mechanism to resist fish predation. For example, the newts, North American Bullfrogs (*Lithobates catesbeianus*), and many toad species contain compounds in their skins that are aversive to many potential predators. Although these are occasionally eaten by game fish, when these predators are faced with a choice, they are avoided. Others have decoys or cryptic colorations that allow them to avoid detection by fish predators. In fact, some escape predation simply by remaining in abodes, secluded from the hungry eyes of a marauding bass. There can be no doubt that much of our amphibian extinction crisis in the United States was predisposed by the loss of breeding sites caused by introduction of highly predaceous fishes. These species were deliberately bred, raised, and introduced around the country, and the fallout for the amphibian community cannot be underestimated.

Among the first fish introduced to North America was the European Carp, *Cyprinus carpio*. These animals were considered a delicacy in Europe and widely eaten. The history of carp introduction extends to the Romans, who spread the species throughout much of Europe. Romans took chariots or wagons and packed them with wet straw, then placed the carp under the moist bedding to avoid desiccation during transport. They could carry carp for miles this way, ensuring a bountiful food source for Roman citizens. The early introductions to North America were for the same reasons, but the impacts could not have been as beneficial for aquatic species as they were for the immediate needs of humans. Carp are large, powerful fish capable of invading breeding sites, destroying native vegetation, and largely making previously ideal breeding habitats unsuitable for fishes and amphibians. Even the novice can recognize damage from introduced carp. Carp are also very prolific and can rapidly overpopulate ponds, leading to water quality issues and exhausting resources for other organisms.

The reproductive potential of carp as a group can be seen very clearly with the introduction of the related Bighead Carp, *Hypophthalmichthys nobilis*, in the Mississippi River basin. This species reaches enormous populations and has catastrophic impacts. This species was introduced for vegetation control and is now an environmental disaster unfolding.

Aquaculture-produced Mosquitofish, *Gambusia affinis*, have been introduced worldwide, although they are actually native to the southeastern United States. Generally, they are thought of as beneficial in that they feed on insect pests, especially disease-harboring mosquitoes. However, larval salamanders and baby turtles undoubtedly used this bountiful food source prior to the introduction of this species. The Mosquitofish is a prolific live-bearing species and can overtake habitats rapidly. In the Arkansas Delta, a visit to many irrigation canals in late fall after an early freeze can find many a shoreline littered with a shocking number of dead Mosquitofish. The influence this has on water quality has received little attention, but it does not take a large imagination to infer that those fish may lead to nutrient pulses during colder temperatures, when the capacity to negate them is at its least. Ammonia toxicity must, at times, impact these artificial aquatic ecosystems and resident species that include many amphibians and reptiles.

There is perhaps no greater catastrophe in relation to aquaculture's impact on the environment than the introduction of non-native species from the tropical fish industry in Florida. The early production of tropical fish in particular has led to mass introductions of almost any species observed in a pet store. Of particular importance was the Walking Catfish. This species is a food and hobby fish that can traverse land areas to invade new water bodies. Once introduced, it prolifically takes over the habitat, eating or otherwise harassing other species until they disappear. At one time, its introduction was considered a threat to the Everglades Kite, due to competition for the bird's preferred food, the Apple Snail. The impact of this species on amphibian and reptile communities is largely unknown.

Much of this discussion has focused on negative impacts due to introductions of exotic fishes that likely negatively affect native herpetofauna. However, the aquaculture industry has undoubtedly contributed to introductions of amphibians and reptiles around the world. One of the most publicized exotic and invasive species is the North American Bullfrog. One could write an encyclopedic volume covering the impacts of North American Bullfrogs on native amphibians, fish, and other fauna. The animals are large, voracious, and exceptionally hardy. North American Bullfrogs have been introduced internationally as a food animal. The same is true in North America; however, here they are commonly transported as biological contaminants in aquaculturally produced fish for stocking. Stocking may be one of the easiest ways for amphibians to be translocated via aquaculture. Several cases of disjunct populations of amphibians can be directly traced to aquaculture. Further, studies demonstrate that in southeastern fish hatcheries where North American Bullfrogs breed, there are high incidences of chytrids, a serious threat to amphibian survival today.

Probably the biggest failure in biological control has been the introduction of exotic invasive Cane Toads (*Rhinella marina*) around the globe. These animals were produced via aquaculture for use as a biological control of beetles that damage sugarcane. Yes, Cane Toads can and will eat sugarcane beetles. However, sugarcane is a very tall plant, and Cane Toads do not climb. Effectively, the entire program was a classic example of not doing your homework. Today, the Cane Toad has invaded sugar-producing and other regions around the globe. The species is very toxic, and there have been increasing records of animals found dead with Cane Toads in their mouths or stomachs. Further, the Cane Toad is an aggressive feeder and very large. It will eat anything up to the size of a small mouse. Reports of them eating snakes, other amphibians, and other animals abound. Finally, the species is extremely prolific and can rapidly grow in population to dominate areas. The introduction of the Cane Toad has been used as a classic example for decades of good intentions gone awry.

The overlap between the commercial and conservation sectors of aquaculture is growing. In Oklahoma, the Oklahoma Department of Wildlife Conservation has even contracted turtle farmers to produce Alligator Snapping Turtles for reintroduction where populations were decimated. Ultimately, if we can produce birds, mammals, and fish suitable for reintroduction, there is really no reason we cannot do the same for herpetofauna. Still, this is something that should be done with great care. The complex systematics and ecological requirements of these species provide unique complications for evaluating habitat as suitable for reintroductions.

The contribution of aquaculture is not all bad. Aquaculture has developed methods for the large-scale production of animals that can be adopted, adapted, and implemented when captive propagation is the last resort. Probably the most well-publicized reintroduction programs for herpetofauna have been those for Sea Turtles in the southern coastal states of North America. These programs were entirely aquacultural in nature. Management of American Alligator populations in Louisiana involves aquaculture producers releasing a percentage of their stock into the wild each year. Thus far, the primary concerns are proliferation of deleterious genotypes in the native population. Many are fearful that these programs could introduce disease to the wild; however, a properly maintained, biosecure facility should produce disease-free animals in the first place. Unfortunately, the reputation of producers, especially in light of the baby turtle–salmonella epidemic of the 1970s, is not one of producing disease-free stock. However, for proper animal husbandry to take place, it is essential to do this very thing. Turtle and frog farms in China are amazing ventures to behold. Although largely serving a culinary end, the animals are housed in very clean facilities. Why the concern for cleanliness in aquatic organisms destined for the plate? It is simple. Animals housed in clean facilities taste better. In India, the Ganges River is notoriously polluted, placing Gharial and Batagurine turtle populations on the brink. There is significant aquaculture activity in this region targeting reintroduction of these species.

Perhaps one of the more controversial elements of herpetofauna aquaculture is the pet industry. Today, we are faced with a commercial demand for herpetofauna as household pets, oddities, or fascinations. Herpetofauna are also in demand for folk medicines in the Far East. The strategy of dealing with this demand has been to suppress trade and educate the citizenry against using these animals; however, economic law suggests that to some degree this might actually increase the value of these species and lead to lucrative black markets. Organized agricultural production seldom involves captive wild animals, but rather leads ultimately to captive stocks raised in captivity over generations. This activity becomes streamlined and the costs drop, leading to reduced prices for products. In fact, there is a reason most of us use a grocery store rather than the wild for food. The store is easier, cheaper, and the product better than what you could get elsewhere. Further, economic law demonstrates that when demand exists, even if it is among a minority of the population, a dropping supply typically leads to higher prices. Consequently, when we reduce the supply of baby turtles, tadpoles, etc. in the pet trade, we definitely raise the value of those products on the black market. Because of this, some conservationists are considering whether elimination of supply lines is in the best interest of the species. By flooding the market with animals, the price should drop, relieving poaching pressure on wild populations. The primary problem with this logic is the role of marketing. In China, wild-sourced products are promoted as more vital than captive-reared organisms. Thus, production could have little effect. However, flooding the market with folk medicines that largely have no medical value may actually lead to the demise of these products as people learn that they simply don't work.

Ultimately, the role of aquaculture was once largely one of producing organisms to serve the interests of the human palate. Today, the profession is a more complex industry that serves many sectors of society and may even prove vital to the very species it once exploited.

Salamanders (Caudata)

Mole Salamanders: Ambystomatidae

Western Tiger Salamander
Ambystoma mavortium Baird, 1850

Identification: The Western Tiger Salamander is a large, up to 8.25 in., ambystomatid salamander. Taxonomic arrangements of this form vary. The California Tiger Salamander, *A. californiense,* was once considered a subspecies of the Eastern Tiger Salamander, *A. tigrinum tigrinum.* The Western Tiger Salamander, *A. mavortium,* was also once considered a subspecies of the Eastern Tiger Salamander. It has also been recognized as a polytypic species with five recognized forms: Barred Tiger Salamander, *A. m. mavortium,* Gray Tiger Salamander, *A. m. diaboli,* Blotched Tiger Salamander, *A. m. melanostictum,* Arizona Tiger Salamander, *A. m. nebulosum,* and Sonoran Tiger Salamander, *A. m. stebbinsi.* In yet other arrangements, the Western Tiger Salamander and all of its subspecies are subsumed into the single monotypic species, *A. tigrinum.* Among the regionally distinct forms, individuals are barred or blotched in sharply or weakly contrasting brown, black, olive, or gray colors with yellow lines, spots, or mottling. Some individuals are strikingly patterned in black and yellow. The color pattern may vary ontogenetically. Males have relatively longer tails than females but are generally a little shorter than females. The cloacae of fertile males are swollen during the breeding season. The Western Tiger Salamander is native to the western and central portion of the United States, with eastern border at approximately the 95th meridian.

Introduction history and geographic range: This species has a long history of being transported around the United States for all the wrong reasons: initially as fish bait, and presently as part of the pet trade. The effects of human-mediated dispersal and establishment of extralimital populations are evident today in hybrid swarming, species replacement, and confused biogeography. The present taxonomic arrangement adds additional concerns of human-mediated mixing of recognized regionally distinct forms of the Western Tiger Salamander. In the United States, the Western Tiger Salamander is known in Arizona, California, Idaho, Nevada, Oregon, and Washington. Problematically, the derivation of these introduced populations confounds clarification taxonomy and historical biogeography.

Ecology: Much of its ecology is variable among populations and even within populations. The introduced sites generally are ponds and reservoirs; however, they require friable soil within which they are generally found much of the year.

The neotenic form requires permanent fishless wetlands. The breeding season varies with latitude and elevation and may last from one to several months. It is a pond-breeder, and depending on the subspecies (or species), eggs are laid singly or in variously shaped masses. Egg masses vary in number but generally comprise 40–50 eggs. Clutch size, however, can vary from an average of about 400 eggs in the terrestrial form to >6,000 eggs in the aquatic large-gilled morph. Eggs hatch in three to seven weeks depending on environmental factors. Larvae can transform in 2.5 months or can overwinter. Transformation size is larger in individuals that metamorphose as sexually mature adults. Neotenic adults may occupy the trophic position of top aquatic predator. Terrestrial adults feed on suitably sized invertebrates and vertebrates. Larvae, especially

Photo by Suzanne Collins.

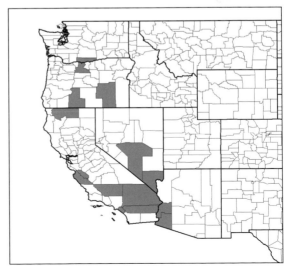

those of the cannibal morph, can eat small vertebrates. Beginning at the egg stage, the Western Tiger Salamander is subject to a wide range of predators during its potentially long life: Eastern Hog-nosed Snake, *Heterodon platirhinos,* Great Blue Heron, *Ardea herodias,* and Raccoon, *Procyon lotor,* to name a few. Extralimital human-mediated dispersal of the Western Tiger Salamander brings unintentional consequences to the survival of congeners. The Western Tiger Salamander will hybridize with the California Tiger Salamander in California (Monterrey and San Benito counties). Chytridiomycosis (chytrid fungus) transmission is also a risk.

Lungless Salamanders: Plethodontidae

SEAL SALAMANDER
Desmognathus monticola Dunn, 1916

Identification: The Seal Salamander is a medium-sized, up to 6.0 in., plethodontid salamander. Dorsal background is gray-brown patterned with darker variably shaped markings, although vermiculations are often reduced. Venter is light gray in adults. Body is stout, and eyes are large. The distal end of the tail is keeled. The Seal Salamander is nearly restricted to the Appalachian Mountains of the East.

Introduction history and geographic range: This species is not native to Arkansas, where it is known from Spavinaw Creek, Benton County. A population was detected in a spring in 2003 and reported in 2004. The colony was derived from northern Georgia in association with the bait market.

Photo by Suzanne Collins.

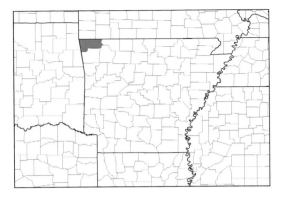

Ecology: Cool, well-aerated rocky streams are the preferred habitat for this stream salamander. It is confined primarily, but not exclusively, to the Appalachian Mountains; however, populations occur in the Southeastern Coastal Plain. In Arkansas, abundances can be high. Recent surveys have detected a more or less continuous population up to 4.4 mi. upstream from the original colony. Two nematodes, *Omea papillocauda* and *O. brimleyorum*, were found in the Benton County population. The mating season is long, and females may breed annually. A few dozen eggs are laid under rocks in the streambed or in the stream bank near flowing water. Hanging individually by stalks, eggs are often attended by the female. Hatching time varies but is typically in late summer to early fall. Larvae transform the following year. Sexual maturity is reached in four to five years of age. This species feeds on invertebrates. No other stream salamander approaching the size of the Seal Salamander occurs on the Ozark Plateau, which could explain its colonization success and likewise raises the question of negative impacts on syntopic salamander species.

Southern Two-lined Salamander
Eurycea cirregera (Green, 1831)

Identification: The Southern Two-lined Salamander is a small, up to 4.0 in., plethodontid salamander. The dorsal background is yellowish. A broad mid-dorsal stripe is flecked in black and bordered on either side by a dark line. Venter is yellow. Body is slender and the tail is long. The Southern Two-lined Salamander is native to much of the southeastern United States

Introduction history and geographic range: The Southern Two-lined Salamander is native to eastern Illinois; however, it is exotic to McKee Creek, Brown County, and LaMoine River, McDonough County, in western Illinois. Individuals were deliberately transported from Fountain, Montgomery, and Parke counties, Indiana, during the 1970s and 1980s.

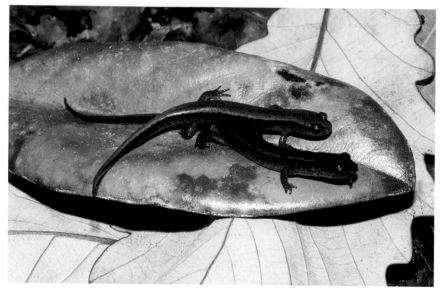

Photo by Suzanne Collins.

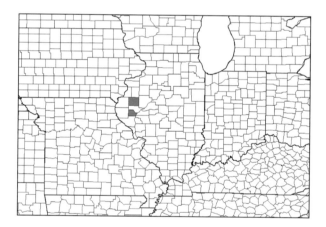

Ecology: This species is typically encountered under cover along margins of rocky streams, but individuals will also wander far from water in forests. Breeding takes place during fall, winter, and spring. Clutch sizes are generally fewer than 100 eggs. Eggs are typically laid singly on the undersides of submerged rocks of streams. The larval period lasts one to three years, and juveniles can reach maturity two years later. Diet of metamorphosed individuals comprises small invertebrates, such as spiders, ticks, beetles, and thrips.

PART 2

FROGS AND TOADS

Preceding the frog and toad species accounts, we present two invited essays. The first essay, by Karen Beard, discusses colonization pathways of the Coqui and potential for future expansion of its exotic geographic range. The second essay, by Brian Gratwicke, discusses detrimental effects of exotic anurans on native biota and how those provide an important justification for the ways frogs matter relative to exotic species impacts.

The Invasive Coqui

How Far Will It Go, and What Can We Do to Stop It?

KAREN H. BEARD

The old adage by Benjamin Franklin—an ounce of prevention is worth a pound of cure—couldn't be truer than when combating the spread of non-native species. With global trade and human travel increasing, the importance of understanding the impact of accidentally introducing animals and intentionally releasing pets needs to come to the forefront. The introduction of the Coqui, *Eleutherodactylus coqui,* to Hawai'i is a case in point. This species is native to Puerto Rico but has completely spread across the Big Island of Hawai'i, where it is primed to spread to new areas. How far will the species go, and what can be done to prevent its future dispersal?

To provide a little background, the Coqui was accidentally introduced to Hawai'i via the nursery trade in the late 1980s, where it joined about 30 other non-native reptiles and amphibians that had already been introduced intentionally, as biocontrol agents or through the pet trade, and accidentally, as stowaways in cargo and nursery plants. Once in Hawai'i, Coquis found a new home with appropriate climate, abundant prey, and many places to breed in, especially lava rock. Coquis have direct development, meaning no tadpole phase, but rather tiny froglets emerge from their eggs, and their populations now reach up to 91,000 frogs/ha, in some places.

Early on in the colonization process of the Coqui, a herpetologist named Fred Kraus warned Hawaiians that this species could become problematic and would be hard to get rid of if they did not act fast, but no one reacted at first. Meanwhile, Coquis spread across the Big Island, mostly through the sale and movement of infested nursery plants, but also through intentional introductions, as some people in Hawai'i enjoyed the frog's call or mistakenly thought it had important insect biocontrol value. About 10 years after the initial introduction, the state of Hawai'i started spending millions of dollars to control and eradicate the frog, mostly because its loud two-note mating call ("c'-kee") (80–90 decibels at 18 inches) disturbed sleepy residents and began to lower property values. The state and local residents tried to control the frogs mostly by removing vegetation and spraying citric acid into infested areas. While these efforts successfully eradicated the frog from O'ahu and Kaua'i, the frog was so widespread on the Big Island that it is now thought to be impossible to eradicate there.

Now that the Coqui has taken up permanent residence in Hawai'i, it will continue to be introduced to "Coqui-free" Hawaiian islands through the nursery trade and inter-island travel. Only with continued vigilance can we prevent its spread to these areas. The establishment of the Coqui in Hawai'i is also relevant to other US states and territories. Coquis from Hawai'i have already been introduced to Guam and California. The introduction to Guam was prevented by port authorities who recognized the frog. Around the same time, the quieter Cuban Flat-headed Frog, *Eleutherodactylus planirostris*, was introduced from Hawai'i to Guam, where it is now widespread.

Over the past decade, the Coqui has been introduced several times to California in nursery plant shipments. The Coqui is now found in and around nurseries in southern California. Recognizing the potential issue, the California Department of Fish and Game placed the Coqui on the state's restricted species list. This requires that the species can be possessed only with a permit, and the restriction provides state and local agencies with regulatory authority to inspect, stop shipments, quarantine, and destroy individual frogs.

There are questions about whether or not Coquis can establish long-term breeding populations outside of nurseries in California. We can gain some insight into this issue from Coqui introductions to other areas in the United States. It never established in Gulf States, like Louisiana, even though it has been introduced there, presumably because the climate is too cold. It is believed that it was in part eradicable from O'ahu because of cold winters. Therefore, it appears that despite some speculation to the contrary, this subtropical species might not be able to establish in some areas where winters are too cool.

Research conducted on the Big Island of Hawai'i suggests that the Coqui will not establish breeding populations above about 4,500 feet because it is too cold. Similar modeling approaches could be used for the continental United States to determine where they are most likely to establish and spread. These models also might want to include potential climate changes, because with warmer and wetter conditions their potential range is likely to expand.

So, how far will the Coqui go? This depends on the behavior of humans. If the goal is for native species to dominate natural habitats, then we have to stop the initial introduction of non-native species. Many non-native amphibians have spread because we intentionally released them. As the Coqui colonization pattern shows, even small populations, like those brought over in a few nursery plants can, under the right circumstances, result in a widespread and abundant exotic species. Preventing initial establishment and eradicating populations that are still small or contained can sometimes be the only option to prevent eventual dispersal.

Context Matters in the Fight to Save Frogs

BRIAN GRATWICKE

As a conservation biologist, I am asked most frequently, "Why do frogs matter?" and it is a deceptively challenging question to answer. It is exasperating, because my personal motivations for working to conserve amphibians aren't easily classified into a rational human values framework. I often talk about biomedical and ecosystem service values of frogs, but these defensive intellectual justifications seldom generate the emotional response needed to be convincing. I am one of those weird people who gets an endorphin rush from seeing a new amphibian species in the wild for the first time. I get a thrill when I experience a cacophony of a thousand frogs of different shapes and colors yelling above one another in a dark jungle pond. My personal testimony is that when the forest loses its nocturnal soundtrack it becomes a poorer place, but my scientific training makes me deeply uncomfortable with expressing personal feelings. Emotions tend to cloud facts and make it difficult to do good science, which requires objectivity.

Nonetheless, it is important to recognize that conservation is essentially a values-based endeavor and that emotional and moral framing of arguments is more likely to influence people in the competitive marketplace of ideas. For example, the "Save the Whales" movement campaigned very successfully in the 1970s on the simple notion that whales are good. They framed the whaling debate as a values conflict between heartless whale hunters and those who could see that whales are obviously gentle, sentient beings. This has proven to be a more challenging conundrum for amphibian conservationists who have struggled to get the "frogs matter" message out to the public. Messages that conflict with the "frogs are good" narrative undermine what little incremental progress we have made in this arena, which is why many of us get heartburn discussing exceptions, such as invasive amphibian species.

One classic example of an invasive amphibian species is the Cane Toad, *Rhinella marina,* introduced to Australia in 1945 as a form of biological control for pests in sugarcane fields. They are highly toxic to predators like snakes, goannas, crocodiles, and quolls, and populations of these predators appear reduced in newly invaded parts of Australia where predators have not yet developed avoidance behaviors. In one area where Cane Toads recently spread, Australian Freshwater Crocodiles, *Crocodylus johnsoni,* ate Cane Toads, causing mass mortalities that reduced the crocodile population by 77%. The detrimental consequences of Cane Toad invasions in Australia have raised similar concerns over the potential effects of Asian Toads, *Duttaphrynus melanostictus,* first detected in 2014 in Madagascar.

The Coqui, *Eleutherodactylus coqui,* is native to Puerto Rico, where it is a cherished animal and important symbol of place and national identity. In Hawai'i, however, it is an invasive species, and in ideal habitats with no native frogs, their populations exploded, forming much higher densities than in Puerto Rico. Homeowners are not enthralled by choruses at 70–80 decibels keeping them up at night, and ecologists worry about the negative ecological effects of these abundant indiscriminate predators of small invertebrates, including endangered snails. Conservationists in Hawai'i are therefore working to eradicate Coquis using a citric acid spray, among other strategies. Back in Puerto Rico the common Coqui is still common, but other Coqui species are in danger of extinction and are the subject of conservation efforts.

In addition to direct impacts of invasive amphibians, some non-native species can have indirect effects on native amphibians by spreading disease. In parts of Korea, invasive North American Bullfrogs, *Lithobates catesbeianus,* have increased prevalence of the amphibian chytrid fungus and reduced populations of endangered Suweon Treefrogs, *Dryophytes suweonensis,* and in Brazil, North American Bullfrogs have been associated with the spread of a deadly global pandemic lineage of the fungus. It is likely that the global trade in amphibians for food or pets has facilitated the spread of the amphibian chytrid fungus, and similar concerns about the potential spread of a newly discovered salamander chytrid fungus have led to an international salamander trade moratorium in the United States. So far, we have not detected the salamander chytrid fungus on any wild or pet salamanders in the United States, but in Europe the fungus has spread rapidly, devastating several wild populations of Fire Salamanders, *Salamandra salamandra.*

As with any non-native species, it is not always possible to predict their effects on native wildlife. In some places, habitat modifications in some cases have facilitated their spread. Asian Bullfrogs, *Hoplobatrachus tigrinus,* thrive in rice paddies in Madagascar, where they are also hunted for food. In Australia, Green and Golden Bell Frogs, *Litoria aurea,* are highly threatened by Mosquitofish introduced from the United States. The fish prey on the frogs' eggs, reducing their populations. The frogs, however, thrive in New Zealand farm ponds despite populations having collapsed in their native range. Habitat requirements of Green and Golden Bell Frogs are quite different from those of native New Zealand frogs, and so they don't harm the native frogs. Some people view the fact that they are doing well there as a good thing, and a review paper on Green and Golden Bell Frogs in New Zealand illustrates the amenity values placed by the researchers themselves on the frogs: "Because these species [Green and Golden Bell Frogs] are not protected in New Zealand, field studies can also incorporate experimental manipulations not readily possible in Australia" (Pyke et al. 2002, Royal Zoological Society of NSW).

It is clear that in the fight to save frogs, context matters. The species, its ecology, the value systems of the people living alongside them, and the audience of your message are all critical. In the case of invasive species, pet owners are a critical target audience that should already be on board with the "frogs matter" message. We have many examples of unintended detrimental consequences from invasive amphibian species, so responsible pet owners should never release non-native species into the wild, because frogs *do* matter.

Frogs and Toads (Anura)

Toads: Bufonidae

CANE TOAD
Rhinella marina (Linnaeus, 1758)

Identification: The Cane Toad is a large, up to 6.0 in., bufonid toad. Among adults, males are smaller than females. Parotoid glands of the Cane Toad are disproportionally larger than those of other North American toads. Their dorsum varies from solid brown to mottled shades of brown and gray. Juveniles and

Photo by Suzanne Collins.

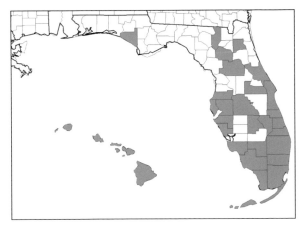

adult females are similar in appearance. Fertile males have cornified thumbs and a dorsum of sandpaper texture and cinnamon color. Its center of distribution is the northern portion of South America.

Introduction history and geographic range: Arrival of this species to the United States was through intentional introduction to Hawai'i, on O'ahu, in the 1930s with the failed expectation that these terrestrial and mostly nocturnal animals would control sugarcane beetles. The grubs of this beetle feed underground on the roots of the plant, and the flying adults eat the leaves. Likewise, for pest control in Florida, 200 individuals were released in Belle Glade and Canal Point, Palm Beach County, prior to 1936, and in Clewiston, Hendry County, prior to 1944. The Cane Toad was reported from an unspecified location in southern Florida in 1957. Intentional pet-related introductions of the Cane Toad in Florida took place in Pembroke Pines, Broward County, in 1963, and in Kendal, Miami-Dade County, in 1964, both by animal dealers. However, it was an escape of approximately 100 toads from an animal importer at the Miami International Airport in Miami, Miami-Dade County, before 1955 and reported in 1966 that is the presumed source of the established population in Florida. It was reported on Stock Island, Monroe County, Florida, in 1970. Subsequent dispersal occurs intentionally and also unintentionally in pots of ornamental plants. In the United States, the Cane Toad occurs nearly continuously through southern and central Florida, and sporadically in northern Florida. It is well established in Hawai'i (Hawai'i, Kau'i, Lāna'i, Moloka'i, Maui, and O'ahu).

Ecology: The Cane Toad is highly adaptable around humans and can thrive in agricultural fields, golf courses, residential developments, and along canals, among other disturbed habitats. Adults are primarily nocturnal and are stimulated to move by rain. Toadlets are distinctly diurnal in their activity. Breeding can occur any time of the year depending on conditions. Calling can be heard throughout the year and sounds much like a very low trill. In Florida, tadpoles are generally seen during March–October. At least two clutches are possible in southern Florida, with the first strong pulse of breeding taking place in late winter. Eggs are laid in rosary-like strings in shallow water of lake, pond, or canal edges, estuarine systems, and puddles and can hatch in a few days. Tadpoles are blackish in color and will swim in schools. Transformation into 0.4 in. toadlets occurs in just over one month in the summer. Individuals are sexually mature within one year of post-metamorphic life.

The Cane Toad eats a wide range of live prey and often will congregate around lighted buildings to capture light-attracted insects. In southern Florida, diet includes the Southern Ring-necked Snake, *Diadophis punctatus punctatus*, Common Ribbonsnake, *Thamnophis sauritus sauritus*, the Brahminy Blind Snake, and especially beetles and ants. Through olfaction, individuals are attracted to such foods as chicken and beef scraps, spareribs, roadkill, dog food, and dog feces. Cars are a major source of mortality of post-metamorphic individuals. Adults are eaten by the Red-shouldered Hawk, *Buteo lineatus,* and toadlets are eaten by the Blue Jay, *Cyanocitta cristata,* and Northern Mockingbird, *Mimus polyglottos.* In Hawai'i, it is preyed upon by the Wattle-necked Softshell.

It remains to be seen whether the Mesoamerican Cane Toad, *R. horribilis,* native to extreme southern Texas and very similar in appearance to the Cane Toad, is also present in Florida.

Poison Dart Frogs: Dendrobatidae

GREEN-AND-BLACK POISON DART FROG
Dendrobates auratus (Girard, 1855)

Identification: The Green-and-black Poison Dart Frog is a medium-sized, 1.4 in., dendrobatid frog. The common name well describes the species, with its strikingly patterned body in green and black similar in both sexes and juveniles. It is a Neotropical species.

"Dendrobates auratus Bocas Del Torro" by brian.gratwicke is licensed under CC BY 2.0. https://www.flickr.com/photos/19731486@N07/6118350016.

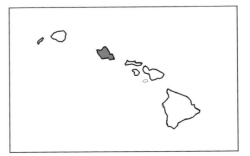

Introduction history and geographic range: Arrival of this species in the United States was through intentional release onto O'ahu, Hawai'i, in the early 1930s for control of non-native insect pests. In the United States, the Green-and-black Poison Dart Frog is restricted to a few valleys on O'ahu, Hawai'i.

Ecology: This species prefers well-shaded forests, where it can be found on the forest floor and aboveground. Its activity is greatest in the morning and after rain. It is least active in the heat of summer. The male fertilizes the eggs after the female lays a 5–7-egg clutch on a moist surface. Upon hatching nearly two weeks later, tadpoles are transported by the male to aquatic sites, such as bromeliad tanks. The tadpoles are carnivorous and transform in slightly more than one month. Individuals are mature by the end of one year or soon thereafter. Ants form the mainstay of its diet. The Green-and-black Poison Dart Frog is thought to be sensitive to habitat disturbance, and its ecological impacts on O'ahu are unknown.

Rainfrogs: Eleutherodactylidae

Coqui

Eleutherodactylus coqui Thomas, 1966

Identification: The Coqui is a small, approximately 1.5–2.25 in., eleutherodactylid frog. Among adults, males are smaller than females. This species is brown to gray-brown in dorsal color and patterned variably. An interorbital bar and a light dorsolateral stripe may be present. Its center of distribution is Puerto Rico.

Introduction history and geographic range: Arrival of this species in the United States was through human-mediated dispersal to Miami, Miami-Dade County,

Photo courtesy of Ryan Choi.

Photo by Suzanne Collins.

Florida, where it was reported in 1975. Extirpated by a freeze in 1977, the Co-qui was subsequently reported from in and around bromeliad greenhouses in southern Miami-Dade County in 1984, where it had been heard since 1976 and was still present in 2000. In the 1990s, this species was introduced to Hawai'i through the nursery trade. Subsequent dispersal readily occurs in ornamental plants, especially bromeliads, which serve as retreats and nesting sites.

Ecology: In the United States, the Coqui is restricted to greenhouses and immediate surroundings in extreme southern mainland Florida and is well established on the Hawaiian Islands of Hawai'i, Kaua'i, Maui, and O'ahu. Adults are terrestrial while hiding during the day and emerge to hunt arboreally at night. Juveniles are terrestrial. More than 9,000 individuals/ha (3,643.7 individuals/acre) have been estimated in Hawai'i. In Hawai'i, breeding can occur each or every other month. In southern Florida, breeding is most common from spring to fall. Males are territorial and aggressively defend clutches of 12–24 eggs laid in moist retreats, such as leaf folds of bromeliads. The tadpole stage is passed in the egg, and hatchlings resemble miniature adults. The male's advertisement call is a very loud version of its common name. Males also have aggressive calls, and females produce a softer call in association with defense of feeding territories. In Hawai'i, adults reach sexual maturity in eight to nine months.

The Coqui feeds primarily on invertebrates but will also eat eggs of conspecifics. Hawaiian populations include native invertebrates in their diet. In Hawai'i, it is preyed upon by a variety of birds and reptiles, as well as by spiders and scorpions. The Small Indian Mongoose, *Herpestes auropunctatus*, is a significant predator. The Cuban Treefrog and the Tokay Gecko are likely significant predators of the Coqui. The Coqui is thought to be a competitor of native birds for food in Hawai'i. Its trophic relationship with other herpetofauna is unknown. Presently, citric acid is the most effective method of control of this species in Hawai'i, where it may be controlled on all islands of Hawai'i. Kaua'i was considered Coqui-free on at least one occasion. Presently, this species may or may not be present on that island.

RIO GRANDE CHIRPING FROG
Eleutherodactylus cystignathoides (Cope, 1877)

Identification: The Rio Grande Chirping Frog is a small, approximately 1.0 in., eleutherodactylid frog. Among adults, males are smaller than females. This species is sandy colored with black stippling on the dorsum. Its rear legs are banded in brown, and it has a black eye stripe. The Rio Grande Chirping Frog is a

Photo by Suzanne Collins.

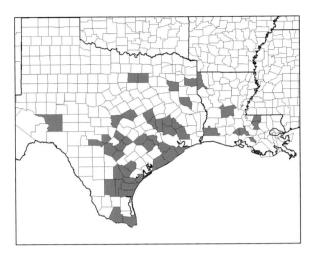

Central American species whose natural geographic range extends northward into Cameron and Hidalgo counties of extreme southern Texas.

Introduction history and geographic range: In the United States, the Rio Grande Chirping Frog is exotic to scattered counties of eastern Texas and in Louisiana. Its extralimital dispersal appears to be through the potted plant industry from the Rio Grande Valley.

Ecology: The species is associated with river alluvium and moist places and around human residences, where it burrows in leaf litter. In Shreveport, males call from early April through late November. The call is a single, low chirp. Between 5 and 13 eggs are laid in moist retreats April–May. The tadpole stage is passed in the egg, and they hatch approximately two weeks later as 0.3 in. miniature versions of the adults. This diminutive frog is an insectivore whose diet includes beetles and spiders. Human-mediated dispersal of the Rio Grande Chirping Frog could place it in contact with the Cliff Chirping Frog, *E. marnockii*, with unknown impacts.

CUBAN FLAT-HEADED FROG, AKA GREENHOUSE FROG
Eleutherodactylus planirostris (Cope, 1862)

Identification: The Cuban Flat-headed Frog, aka Greenhouse Frog, is a small, approximately 1.25 in., eleutherodactylid frog. This species is brownish in dorsal color. Although some populations may be overwhelmingly striped, most individuals are mottled. Its center of distribution is Cuba.

Introduction history and geographic range: Arrival of this species in the United States was through dispersal to southern Florida, reported in 1875, and to Key West, Monroe County, reported in 1889. In Alabama, it was first reported

Photo by Suzanne Collins.

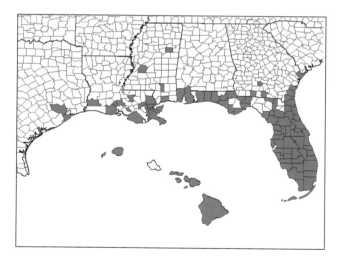

from Fairhope, Baldwin County, in 1983, where it was first found in 1982. As of 2014, the population was still present and had expanded in this coastal county. In Georgia, it was first reported from Savannah, Chatham County, in 1998. In Hawai'i, it was first reported from Hawai'i and O'ahu, in 1999, where it had been introduced through the plant trade since the 1990s. In Louisiana, it was first reported from New Orleans, Orleans Parish, in 1994, with observations since the 1970s. In Mississippi, it was first reported from Gulfport, Harrison County, in 2004, where it was caught in 2003, and in a record from Starkville, Oktibbeha County. Records from Hinds County are indicative of established populations in Jackson. In Texas, it was first reported from La Marque, Galveston County, in 2007. Subsequent dispersal can be highly successful when it is in association with ornamental plants. In some cases, the dispersal through plants is intentional. In the United States, the Cuban Flat-headed Frog is found statewide in Florida, the extreme Southeast, and on the Hawaiian Islands of Hawai'i, Kaua'i, Lāna'i, Maui, and O'ahu.

Ecology: The Cuban Flat-headed Frog occurs in a wide range of habitats, including sandy uplands, especially those that are not too frequently burned, mesophytic forests, and urban gardens. It avoids very wet habitats. Individuals will hide under surface cover and use natural burrows, such as those of the Gopher Tortoise, *Gopherus polyphemus,* where they can be the dominant commensal vertebrate species. Occasionally, individuals are found above the ground in moist confines of vegetation, but it is otherwise terrestrial. This diminutive frog can be very abundant in natural and disturbed sites, and in southern Florida it excels in the aftermath of a hurricane. In Hawai'i, it is most often associated with human-disturbed lowland habitat. Eggs are laid through the summer in the moist retreats. The male's call resembles a soft chirping and can be heard night or day, often after rainfall, or even when watering garden plants among which they reside. The young pass the tadpole stage in the egg and hatch as miniature versions of the adults. Adults are presumed to reach adulthood within one year of life. This species eats invertebrates, especially ants. This finding holds true in Hawai'i, where ants are non-native and negatively impact native invertebrate species. At population densities of up to 12,500 frogs/ha, up to 129,000 invertebrates/ha (52,226.7 invertebrates/acre) can be eaten each night. It is depredated by the Ring-necked Snake in Florida and presumably by the Cuban Treefrog as well, which is a predator of this frog in the West Indies. Impacts on other semifossorial insectivores remain unknown.

Treefrogs: Hylidae

GREEN TREEFROG
Hyla cinerea (Schneider, 1799)

Identification: The Green Treefrog is a small, up to 2.0 in., hylid frog. Among adults, males are smaller than females. Background color is light green but can change to olive or nearly brown. Small bright orange or yellow dorsal spots may be present. A white lateral stripe, variable in length, is usually present. The venter is white. Its center of distribution is the United States, where it occurs along the Southeastern Coastal Plain.

Introduction history and geographic range: In the United States, the Green Treefrog is exotic to Shepherd of the Hills Conservation Center, Taney County,

Photo courtesy of Janson Jones.

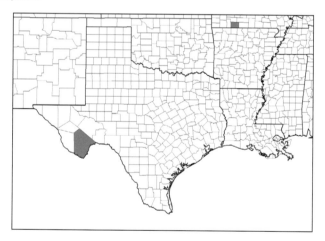

Missouri, where it was first reported in 2012. This species is also exotic to Big Bend National Park, Brewster County, Texas, where it was first reported in 2007.

Ecology: The Green Treefrog is a semiaquatic species and associated with sunny, open, permanent still bodies of water or long-hydroperiod wetlands that have extensive emergent vegetation. It also inhabits adjoining forested upland connections. In Texas, it is abundant in a beaver pond at Rio Grande Village. Most activity is nocturnal, although males will often call in sultry weather and in advance of storms. Breeding season is longest in the South (May–October) and abbreviated in northern populations (May–July). The call sounds much like "bonk." In large breeding congresses, choruses can be wonderfully deafening. An average of about 2,000 eggs are laid in the form of partial clutches. Life history traits vary among locations. Generally speaking, however, eggs hatch in two to three days, and the larval period lasts approximately two months. Sexual maturity can be reached in one or two years after transformation at around 1.75 in.

Across its geographic range the Green Treefrog eats a wide range of small arthropods, especially spiders and beetles. At Big Bend, roaches, beetles, grasshoppers, and crickets were eaten most frequently. Females ate slightly larger prey than males and most of the same prey as males. The dietary breadth of this population was similar to that of most native populations. Dietary overlap of the Green Treefrog is greatest with that of the Squirrel Treefrog in the Everglades. The Green Treefrog is eaten by a wide range of native species, including birds and snakes. It is also readily eaten by the Cuban Treefrog. The Big Bend population may be susceptible to predation by the North American Bullfrog.

SQUIRREL TREEFROG
Hyla squirella Bosc, 1800

Identification: The Squirrel Treefrog is a small, up to 1.5 in., hylid frog. Among adults, males are smaller than females. Dorsal color and pattern are highly variable. Background color is usually light green but can change to various shades of green and brown in metachrosis. Blotches of green or brown of various sizes,

Photo courtesy of Janson Jones.

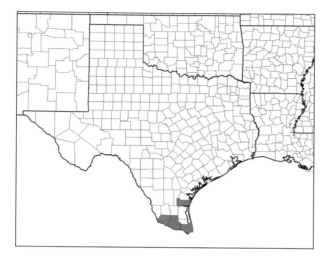

numbers, and shapes may be present on the dorsum. The venter is white. The Squirrel Treefrog is native to the Southeastern Coastal Plain of the United States.

Introduction history and geographic range: In the United States, the Squirrel Treefrog is exotic to Texas (Cameron, Hidalgo, Kleberg, and Starr counties), having been detected in Bentsen–Rio Grande Valley State Park.

Ecology: The Squirrel Treefrog is a semiaquatic species and associated with sunny, open, short- and long-hydroperiod wetlands that have extensive emergent vegetation. It also inhabits adjoining forested upland connections. Most activity is nocturnal, although males will often call in sultry weather and in advance of storms. Breeding season is generally from March through October and varies among years and locations. Winter breeding is known for Florida, and farther north breeding typically occurs from April to August. The call sounds much like "rek." An average of about 1,000 eggs are laid in the form of partial clutches. Life history traits vary among locations. Generally speaking, however, eggs hatch in two to three days, and the larval period lasts for slightly more than one month. Sexual maturity can be reached in one or two years after transformation at around 1.0 in. Flies and beetles dominate an otherwise broad diet of small invertebrates in the Everglades. Dietary overlap is greatest with the Green Treefrog in the Everglades. The Squirrel Treefrog is eaten by a wide range of native species, including birds and snakes. It is also readily eaten by the Cuban Treefrog.

Cuban Treefrog

Osteopilus septentrionalis (Duméril and Bibron, 1841)

Identification: The Cuban Treefrog, or Rana platanera, is a large, up to 6.5 in., hylid frog. Among adults, males are smaller than females. Very large individuals are consistently associated with sites under carrying capacity, such as new colonies or those with poor reproductive success. This species is easily identified by a co-ossified skull, giving it a sandpaper-like texture, and by its distinctively green bones most easily seen in the rear legs from the ventral aspect. Dorsal

Photo by Janson Jones.

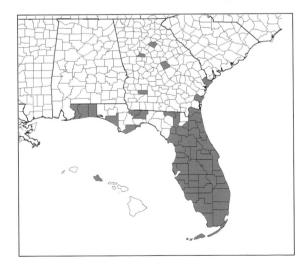

color is highly variable and can fade and darken in color. Most individuals are beige in background and irregularly patterned in darker shades of brown. Some individuals are anteriorly or entirely green. Venter is white. Fertile males can be distinguished by nuptial pads on the thumbs. Its center of distribution is Cuba.

Introduction history and geographic range: The Cuban Treefrog was first reported in the United States from Key West, Florida, in 1931, where it may or may not have been native. First record of the Cuban Treefrog on the Florida mainland was in Miami, Miami-Dade County, in 1952. It was probably absent from the Florida mainland until 1945 and, likely, exotic to mainland Florida. This species was first reported from Georgia in Chatham County in 2004 and published in 2007. In 2019, it was reported as established on Jekyll Island, Glynn County, since 2012. It was first reported in Hawai'i from O'ahu, where southern Florida stock was intentionally released in the 1980s. Subsequent dispersal occurs easily on vehicles and in ornamental plants. In the United States, the Cuban Treefrog occurs nearly continuously through southern and central Florida, and sporadically in northern Florida. It is well established on the Hawaiian island of O'ahu. In time, it is expected to disperse along the Gulf Coast into Texas.

Ecology: The Cuban Treefrog is associated with mesophytic forests, mangroves, and human dwellings. Palm nurseries, banana plantations, and greenhouses can provide ideal conditions for its success and dispersal. Individuals can be found from close to the ground to the tops of trees as well as on human-made structures. Their abundance can be strongly limited by number of refuges. Primarily nocturnal, large individuals will rest outside their retreats in sultry weather and will bask occasionally as well.

Breeding can occur anytime of the year when certain environmental conditions are met. Typically breeding will occur during April–October in much of Florida and April–November in Honolulu. More than one clutch can be produced each year. Calling can be heard throughout the year and sounds much like a grating squawk and may be followed by a series of clicks. In large breeding congresses, calling can be deafening. Hurricanes incite very strong breeding responses. An average of about 4,000 eggs are laid in a surface film and can hatch in just over one day. In the summer, tadpoles transform in less than one month at approximately 0.5 in. Individuals are sexually mature within one year of postmetamorphic life. In the Everglades, most males are dead shortly after one year of postmetamorphic life. Few females live beyond three years. The Cuban Treefrog eats a wide range of invertebrates and small vertebrates, depending on what is available. Roaches, even very large ones, and beetles are especially relished. Frogs, including other Cuban Treefrogs, small snakes, and geckos are eaten as well. Its skill at eating treefrogs negatively impacts the Green Treefrog and the Squirrel Treefrog in natural habitats. On buildings and in areas where the two species are already compromised, colonization by the Cuban Treefrog can result in their extirpation. Adult body sizes of the Indo-Pacific Gecko are larger in the presence of the Cuban Treefrog, presumably the result of more food for fewer geckos and some level of protection associated with larger body size. The Cuban Treefrog is eaten by a wide range of native species, including birds and snakes, and is actively hunted by the Cuban Knight Anole and the Tokay Gecko.

Clawed Frogs: Pipidae

African Clawed Frog
Xenopus laevis (Daudin, 1802)

Identification: The African Clawed Frog is a small pipid frog that can reach 5.5 in. Among adults, males are smaller than females. Individuals have protruding, upward-pointing eyes set close to the snout. The dorsum is darkly vermiculated over a variable tan to brown background color. The venter may or may not be immaculate. The hind legs are long and completely webbed. Males can be discerned by roughened areas on ventral surfaces of forelimbs, and females by enlarged cloacal lips. Its center of distribution is southern Africa.

Introduction history and geographic range: In the United States, the African Clawed Frog was first discovered in Orange County, California, in 1968. The source of introduction in southern California was through release or escape of pets and laboratory stocks associated with pregnancy tests. Subsequent dispersal has been along waterways and by deliberate introductions. In Arizona, it was first reported from Tucson, Pima County, in 1996, where it had been introduced

Photo courtesy of Carlos Nieves.

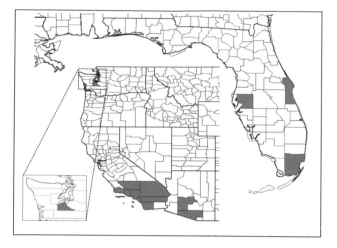

into waterways of a golf course. In Florida, it was first reported from Riverview in Tampa, Hillsborough County, in 1996, and thought to have been collected in the 1970s. It was reported in 2016, and collections made in 2013 confirmed establishment of this species in abandoned aquaculture ponds in Hillsborough County, perhaps since the 1970s. In Washington State, it was first reported in 2017 from Lacy, Thurston County, where it was found to be well established in a settling pond. In the United States, the African Clawed Frog is established in Arizona, streams and ponds of southern California, Florida, and Washington.

Ecology: Although aquatic, individuals will move about land if conditions are suitably wet. They will also aestivate face-up in cavities of their making in drying ponds. This species can tolerate brackish conditions. Very fast-moving water and ponds that ice over completely appear to be unsuitable to its colonization. Breeding occurs during January–November and especially during March–June. Clutches can be laid at monthly intervals during this time. Females can carry up to 17,000 eggs. Presumably, up to about 3,000 eggs are laid in clusters of up to 10 eggs during a reproductive event. Peak calling is during April–May. The call, which can last almost half a minute, is buzz-like and made under the water. Eggs hatch in two or three days. The tadpole is distinctive, with a broad, flat head and tentacles on each side and resembles a miniature and transparent Flathead Catfish, *Pylodictis olivaris*. Tadpoles can be seen feeding in large schools in midwater. In captivity, tadpoles transform in about two months. In Florida, tadpoles were collected in July. Adults generally eat slow-moving aquatic invertebrates. They feed aggressively and, with the use of their hands, can handle large prey. Groups of individuals will grab a large meal with their toothed jaws and render it with their sharp hind claws. Conspecific tadpoles and fish eggs are also eaten. Juveniles of the Western Toad, *Anaxyrus boreas,* are also prey of this species. The African Clawed Frog is subject to predation by many species of centrarchid fish, the Two-striped Gartersnake, *Thamnophis hammondii,* herons, egrets, Common Ravens, *Corvus corax,* and Western Gulls, *Larus occidentalis.* Climate modeling predicts that west Florida will be too warm for this species; however, shade and deep water may provide refuge for some populations.

True Frogs: Ranidae

Japanese Wrinkled Frog
Glandirana rugosa (Temminck and Schlegel, 1838)

Identification: The Japanese Wrinkled Frog is a small, 2.1 in., ranid frog. Among adults, males are smaller than females. The dorsum is dark gray or green. Both the dorsum and legs have many parallel folds of skin, giving the frog an overall wrinkled appearance. The venter is mottled gray with a slight yellow cast. The rear legs are banded, and the rear feet are strongly webbed. Fertile males can be distinguished by nuptial pads on the thumbs. The Japanese Wrinkled Frog is a species of Japan, Korea, and parts of China and the Russian Far East.

Introduction history and geographic range: In the United States, the Japanese Wrinkled Frog was first detected in Hawai'i, where it had been intentionally released on O'ahu in the late 1890s to control introduced insects. In the United States, the Japanese Wrinkled Frog occurs on the islands of Hawai'i, Kaua'i, Maui, and O'ahu. This frog would likely colonize the remaining islands if introduced onto them.

Photo courtesy of Ronald Altig.

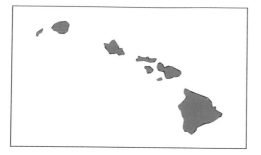

Ecology: The Japanese Wrinkled Frog inhabits wetlands and the quieter portions of cool clear streams at least to mid elevations. Although most often associated with watercourses, individuals will wander well away from water in wet forests of Hawai'i. Breeding occurs during May–September in their native range and during February–August in Hawai'i. Buzzing best describes the sound of the male's courtship call. Clutches of approximately 1,000 eggs are laid in several small clusters usually attached to emergent structures. More than one clutch can be produced each year. Eggs hatch in under one week, and tadpoles probably transform within the year they are laid. Upon larval transformation, froglets can be as large as 1.1 in. The Japanese Wrinkled Frog eats a wide range of prey, especially ants, and will also eat conspecifics.

Rio Grande Leopard Frog

Lithobates berlandieri (Baird, 1959)

Identification: The Rio Grande Leopard Frog is a medium-sized, up to 4.5 in., ranid frog. Among adults, males are smaller than females. The dorsal background color ranges in shades of tan to light olive. Large dark spots are present on the dorsum and sides, and the legs are banded. Dorsolateral folds are distinctive. They are replaced posteriorly with a pair that is inset. The throat is mottled, and the venter is white. Rear feet are webbed. The Rio Grande Leopard Frog is native to southwestern Texas southward into southeastern Mexico.

Introduction history and geographic range: In the United States, the Rio Grande Leopard Frog is exotic to southern portions of Arizona and California of the Colorado River and Imperial Valley.

Ecology: The Rio Grande Leopard Frog is associated most with wet open habitats near moving or still water, often canals, ditches, and rivers. Breeding can occur nearly year-round and will vary among regions and years; however, in Texas, most egg-laying occurs in spring and fall. Calling may be heard throughout most of the year. The Rio Grande Leopard Frog is a generally cooler weather breeder, and calling takes place more often late at night during the hottest times

"Rio Grande Leopard Frog (Lithobates berlandieri), Hwy 4., Cameron Co., TX, USA (25.9442°N, 97.3533°W, 3 m. elev.), 10 April 2016" by William L. Farr is licensed under CC BY-SA 4.0. https://commons.wikimedia.org/w/index.php?curid=90929696.

of the year. Two calls have been identified in males. The advertisement call is a trill. In response, males may make a territorial chuckling call. Eggs are laid in a globular mass in lentic systems or quieter portions of streams. The larval period ranges from four to nine months in duration and depends on when the eggs are laid. The Rio Grande Leopard Frog eats invertebrates and occasionally other frogs. Various birds and mammals will prey upon this species. The Checkered Gartersnake, *Thamnophis marcianus*, is a predator of tadpoles and postmetamorphic individuals. The North American Bullfrog is considered a potential predatory threat to the Rio Grande Leopard Frog.

North American Bullfrog
Lithobates catesbeianus (Shaw, 1802)

Identification: The North American Bullfrog is a large, up to 6.0 in., ranid frog. Among adults, males are smaller than females, and body sizes of both can vary substantially among sites and years. The dorsal color is generally some shade of dull green to brown, and snouts may be bright green. The legs are banded. A distinctive dorsolateral fold curves around the tympanum. The venter is white; however, the dorsum or venter may be mottled or vermiculated in pattern. The rear feet are extensively webbed. Adult males are easily recognized by tympana that are larger in diameter than the eye. Fertile males can be distinguished by enlarged forearms and thumbs and a yellow throat. A southern frog, the North American Bullfrog is native to much of the eastern United States.

Introduction history and geographic range: In the United States, the North American Bullfrog is exotic to Arizona, California, Colorado, Florida, Hawai'i (Hawai'i, Kaho'olawe, Kau'i, Lāna'i, Maui, Moloka'i, and O'ahu), Idaho, Iowa, Minnesota, Montana, Nevada, New Mexico, Oregon, Utah, Washington, and Wyoming. In Florida, it is known from Highlands County, where it was introduced in Hicoria in the mid-1900s for farming. In Hawai'i, it was introduced in the late 1890s for food and exotic insect control from California, where it is also exotic. In Iowa, the North American Bullfrog is native along the Mississippi River and in the southern part of the state; however, it was introduced to Boone, Greene, and Polk counties in the 1940s for human consumption.

Photo courtesy of Eugene Wingert.

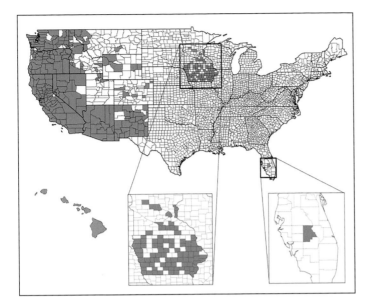

Elsewhere on the mainland, the North American Bullfrog is exotic west of the Rocky Mountains.

Ecology: The North American Bullfrog prefers lentic permanent bodies of water. Northern populations go dormant in winter. In warmer parts of the years, it is active day and night. Preferring a well-vegetated shoreline, it will sit and wait for invertebrate and vertebrate prey and bask as needed to keep warm. Individuals will quickly colonize artificial ponds, almost 291 individuals/acre in a small Arizona pond. Tadpoles, juveniles, and adults of this species have been and continue to be moved around intentionally and unintentionally in the agency of humans. Individuals will quickly colonize artificial ponds. Tadpoles purchased from garden centers for ornamental ponds and those purchased elsewhere for bait carry the risk of transmitting disease, such as chytrid fungus, as well as potentially upsetting the genetic structure of the local population. Breeding season ranges from June–July to March–November, depending on latitude and local conditions. The male's call is a distinctive low-pitched sounding "brrrrum." During the breeding season, adult males may fight with one another. Up to 20,000 eggs are laid in large surface films of up to 40 in. across. The eggs hatch in a few days, and the larval period can last from six months to three years depending on the temperature of the water. The tadpoles can grow to large sizes; the dorsum is generally some shade of dull green and flecked in black. The presence of North American Bullfrog tadpoles can result in prolonged larval times, decreased survivorship, and reduced body mass at transformation of various native anuran species.

The diet of the North American Bullfrog is legendary. From earthworms to mink and small American Alligators in the diet, the North American Bullfrog is a true dietary generalist—a wide trophic breadth to go with a wide mouth! In turn, from eggs to large adults, the North American Bullfrog is subject to a wide range of predators. In Hawai'i, it is preyed upon by the Wattle-necked Softshell turtle. Interestingly, bullfrog tadpoles are unpalatable to fish, which provides a strong colonization advantage to this species. Introduced populations of the North American Bullfrog negatively impact native ranid frogs of Arizona, California, Colorado, Montana, Nevada, and Oregon, native amphibians in Iowa, and the Mexican Gartersnake, *Thamnophis eques*, in Arizona. The North American Bullfrog is considered a potential predatory threat to the Rio Grande Leopard Frog. Its presence is often associated with a decrease in numbers of the North American Green Frog.

NORTH AMERICAN GREEN FROG

Lithobates clamitans (Latreille, 1801)

Identification: The North American Green Frog is a medium-sized, up to 3.0 in., ranid frog. Among adults, males are smaller than females, although the difference is not as pronounced in northern populations. Dorsum is bronze or tan without dorsal spots in the Bronze Frog, *L. c. clamitans*. Dorsum ranges from bright green to dull olive to olive-brown with scattered black spots in the North American Green Frog, *L. c. melanotus*. Adult males are easily recognized

Photo courtesy of Eugene Wingert.

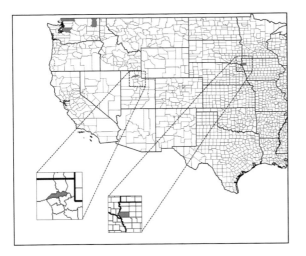

by tympana that are larger in diameter than the eye. Fertile males can be distinguished by enlarged forearms and thumbs and a yellow or yellowish throat. The North American Green Frog is native to much of the eastern United States.

Introduction history and geographic range: In the United States, the North American Green Frog is exotic to Iowa, Utah, and Washington, where it was introduced for human consumption.

Ecology: The North American Green Frog favors open permanent lentic bodies of water with a littoral zone and vegetation. In the South, it may often be found in more wooded habitat. Northerly populations rely on permanent ponds to accommodate overwintering by tadpoles. Northern populations go dormant in winter. In the warmer seasons, individuals are active day and night. Breeding occurs in the summer, earlier and later in the South. The call of the male is reminiscent of the pluck of a banjo string. During the breeding season, adult males may fight with one another. A thin surface film containing up to about 4,000 eggs is laid in shallow, vegetated, generally still water. Eggs will hatch within two weeks, depending on incubation temperatures. The larval period lasts approximately three months; however, tadpoles overwinter in northern regions. Age at maturity varies geographically. Age since metamorphosis to sexual maturity ranges one to two years in males and two to three years in females. This species eats invertebrates and small vertebrates, including other frogs. Individuals are eaten by a wide range of vertebrate and invertebrate predators. Presence of the North American Bullfrog often results in lower abundances of the North American Green Frog.

NORTHERN LEOPARD FROG
Lithobates pipiens (Schreber, 1782)

Identification: The Northern Leopard Frog is a medium-sized, up to 3.5 in., ranid frog. Among adults, males are smaller than females. The dorsal background color ranges through shades of green and sometimes brown. Large dark spots occur in pairs between continuous dorsolateral ridges. Spots are also on the sides. The rear legs are banded. The venter is white. Rear feet are webbed. Fertile males can be distinguished by nuptial pads on the thumbs. The Northern Leopard Frog is native to much of the northern and interior western United States.

Photo courtesy of Eugene Wingert.

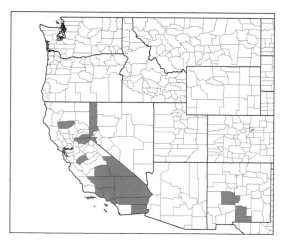

Introduction history and geographic range: In the United States, the Northern Leopard Frog is exotic to California, Nevada, and New Mexico. Native and exotic stock are difficult to differentiate. It was introduced to California for human consumption. In Nevada, it is found in Douglas and Washoe counties in association with Lake Tahoe. In New Mexico, it is established in Otero (Mescalero in Sacramento Mountains) and Socorro (in the Rio Grande, south of the Caballo Reservoir) counties.

Ecology: Once known as the Meadow Frog, this species prefers wet grassy meadows. In some places, early successional fields near breeding ponds can be readily colonized. Breeding occurs in spring or fall or both. A snore-like call is made by the male. A globular egg mass containing up to about 8,000 eggs is laid in shallow, vegetated, generally still water. Eggs hatch within two weeks depending on incubation temperatures. The larval period lasts approximately three months; however, tadpoles are capable of overwintering under certain conditions. Age at maturity varies geographically. Age since metamorphosis to sexual maturity ranges one to two years in males and two to three years in females. The Northern Leopard Frog eats invertebrates and occasionally small vertebrates, including other frogs and conspecifics. Crayfish and gamefish can decimate local populations of the Northern Leopard Frog. Especially in areas where postreproductive individuals remain close to water, the North American Bullfrog can cause severe declines. Tadpoles are subject to heavy predation by the Terrestrial Gartersnake, *Thamnophis elegans,* and the Common Gartersnake, *T. sirtalis.*

SOUTHERN LEOPARD FROG
Lithobates sphenocephalus (Cope, 1886)

Identification: The Southern Leopard Frog is a medium-sized, up to 3.5 in., ranid frog. Among adults, males are smaller than females. The tympanum is dark with a single white spot in its center. The dorsal background color ranges in shades of green and sometimes brown. A few dark spots occur in pairs between dorsolateral ridges. A few spots are also on the sides. The rear legs are banded. The venter is white. Rear feet are webbed. Fertile males can be distinguished by nuptial pads on the thumbs. The Southern Leopard Frog is native to much of the southern and eastern United States.

Introduction history and geographic range: In the United States, the Southern Leopard Frog is exotic to the Santa Ana River and Prado Flood Control Basin in Los Angeles, Orange, Riverside, and San Bernardino counties of California. It has been in Sacramento County since 1992 and in Riverside County, California, where it was introduced with a shipment of North American Bullfrogs, crayfish, and fish from Louisiana in the 1990s. In New York, it is established in Seneca County in association with a military installation.

Photo by Suzanne Collins.

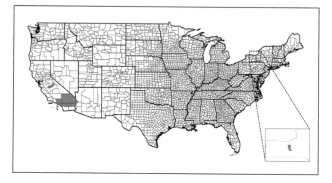

Ecology: The Southern Leopard Frog can be present in a wide range of wetland habitats with a mesic upland connection. It can also be found in coastal areas associated with brackish water. Breeding occurs in the fall or spring in the North and essentially year-round in the South. A halting, chuckling call is made by the male. A globular egg mass containing about 2,000–5,000 eggs is laid in shallow, vegetated, generally still water. Eggs will hatch within three to five days, depending on incubation temperatures. The larval period lasts approximately three months; however, tadpoles are capable of overwintering under certain conditions, such that the larval period could last nine months. Age at maturity is less than one year after larval transformation. The Southern Leopard Frog eats invertebrates and small vertebrates, occasionally other frogs, including conspecifics. It is in turn subject to predation by a wide range of birds, mammals, and snakes.

Wood Frog

Lithobates sylvaticus (LeConte, 1825)

Identification: The Wood Frog is a small-sized, up to 3.0 in., ranid frog. Among adults, males are smaller than females. The dorsal background color ranges from tan to shades of rusty red or brown. A dark mask covers the eyes. The rear legs are banded. The venter is white. Rear feet are webbed. Fertile males can be distinguished by enlarged thumbs. The Wood Frog is native to much of the eastern United States and parts of the Southeast and the West.

Introduction history and geographic range: In the East, the Wood Frog is native to portions of eastern and southeastern Illinois and spottily in adjoining states of Missouri and Indiana; however, it is exotic to McDonough County in

Photo courtesy of Eugene Wingert.

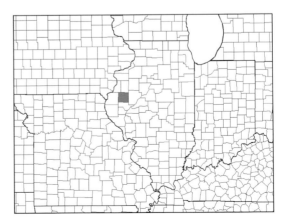

western Illinois, where it was deliberately introduced from Brown and Parkland counties, Indiana, during the 1980s.

Ecology: Nonbreeding individuals are found in forests. Breeding occurs in cooler months, the timing of which varies geographically, with the southernmost populations beginning earliest in the season. The breeding call is akin to the quacking of a duck. A floating globular egg mass containing a few hundred eggs is laid in shallow, fishless pools. Egg masses are characteristically bunched together by females. Eggs hatch in two to three weeks depending on location. Larval transformation occurs in approximately 90 days. Males generally reach maturity one year after larval transformation, and females reach maturity two years after larval transformation. The Wood Frog feeds on invertebrates. Eggs and tadpoles are subject to the depredations of newts, *Notophthalmus* spp., and postmetamorphic individuals are eaten by a wide range of birds, mammals, and snakes.

NORTHERN RED-LEGGED FROG
Rana aurora Baird and Girard, 1853

Identification: The Northern Red-legged Frog is a medium-sized, up to 5.0 in., ranid frog. Among adults, males are smaller than females. The dorsal background

Photo by Suzanne Collins.

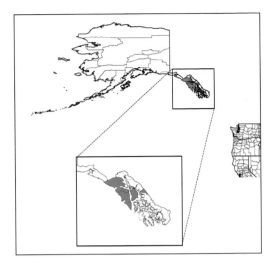

color may be reddish, brown, or gray. Dark spots with a white center are present on the dorsum. A whitish stripe runs along the jaw. The Northern Red-legged Frog is native to the northern half of the West Coast of the United States.

Introduction history and geographic range: In the United States, the Northern Red-legged Frog is exotic to Hoonah on Chichagof Island, along the Alexander Archipelago of Alaska, where it was intentionally introduced from egg clutches that were derived from Oregon in about 1982.

Ecology: Nonbreeding individuals are found in riparian habitat. Males call from under the water, a sound that does not carry far to human ears. Partial clutches of a few hundred eggs each, totaling around 1,000 eggs, are laid in the winter in shallow water. Eggs hatch as late as nine months after deposition, and tadpoles transform in about three months. Males can mature in the season following their birth; however, most mate after two years of age, females one year later than males. It feeds on invertebrates and small vertebrates.

CALIFORNIA RED-LEGGED FROG
Rana draytonii Baird and Girard, 1852

Identification: The California Red-legged Frog is a medium-sized, up to 4.25 in., ranid frog. Among adults, males are smaller than females. The dorsal background color may be reddish, brown, or gray. Solid dark spots are present on the dorsum. A whitish stripe runs along the jaw. Fertile males can be distinguished by enlarged thumbs. The California Red-legged Frog is native to the southern half of the West Coast of the United States.

Introduction history and geographic range: In the United States, the California Red-legged Frog is exotic to Elko, Nye, and White Plains counties, Nevada.

Ecology: Nonbreeding individuals of the California Red-legged frog are found at breeding sites as well as nonbreeding sites that are cool and wet. Males emit a call of three to seven notes from the water surface, a sound that does not carry far to human ears. Clutch sizes average 2,000 eggs laid in parcels that are attached to vegetation near the water surface. Timing of oviposition varies from fall through spring depending on location. The larval period lasts approximately three to seven months, although overwintering is possible in some locations. Sexual maturity is reached at two years in males and three years in females. This species eats invertebrates and vertebrates, with other frogs often making up much of its diet.

Photo courtesy of Daniel Hughes.

PART 3

TURTLES

Before the species accounts, we present an invited essay by R. Bruce Bury and Brent Matsuda, who discuss extralimital colonization of the Pacific Northwest by turtles that are native to the eastern United States.

Introduced and Extralimital Species of Freshwater Turtles in the Pacific Northwest

R. Bruce Bury and Brent M. Matsuda

There are two native turtles in the Pacific Northwest. Western Pond Turtles (*Actinemys marmorata*) occur in western Washington and Oregon, and south to Baja California (Bury et al. 2012). In the Vancouver area, British Columbia (BC), they are considered extirpated, and it is not clear whether they ever were native or introduced (see Carl 1968; Gregory and Campbell 1984; Cook et al. 2005; Matsuda et al. 2006). The Western Painted Turtle (*Chrysemys picta belli*) ranges from Vancouver Island and adjacent mainland in BC, and south to the lower parts of the Willamette Valley, Oregon, and then east.

Today, many introduced or extralimital species of turtles are present (figure 1). Some records are based on museum specimens, while others are reported in the literature (Holland 1994; Bury 2008) or in online reports such as *Nonindigenous Aquatic Species* (USGS 2018). Also, our discussion includes observations and captures from selected reliable unpublished sources (e.g., Washington and Oregon Departments of Fish and Wildlife; turtle researchers in British Columbia).

Most accounts are of Red-eared Slider (*Trachemys scripta elegans)* with increasing reports of Common Snapping Turtle (*Chelydra serpentina*) (figure 2), and a wide variety of other species occasionally recorded (Bury 1995b): emydid "basking" turtles (e.g., map turtles, *Graptemys* spp.), mud/musk turtles (*Kinosternon* spp.), and Spiny Softshell (*Apalone spinifera*) that occur mostly in eastern North America. There are sometimes Asian species (e.g., Reeve's Pond Turtle, *Mauremys reevesii*). The occurrence of both Common Snapping Turtles and Spiny Softshells is often overlooked because they are highly aquatic and seldom surface to engage in atmospheric basking. Females leave the water to nest on land (figure 3), but this may be for only one or two trips per year and often under darkness in the evening or at night.

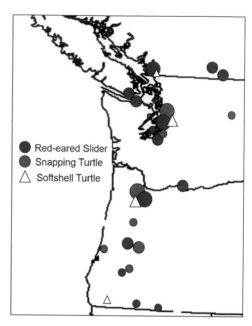

Figure 1. Records and reports of extralimital species of freshwater turtles in the Pacific Northwest. The size of the symbol represents the relative size of the population.

Most of the non-native turtles found in the Pacific Northwest are in urban waters or lowlands. For example, 83% of the turtles found in urban waters were introduced or extralimital, while more open or countryside sites had 1% or less (see Bury 2008). In urban ponds of Eugene, Oregon, a trapping session yielded 37 native Western Pond Turtles and 102 introduced Red-eared Sliders (W. Castillo, Oregon Dept. Fish and Wildlife, unpublished report 2005). In the Greater Vancouver–Fraser Valley area of BC, 232 basking surveys of 70 sites conducted over four years recorded Painted Turtles in 39% of the sites, whereas 85% of the sites had Red-eared Sliders. Sliders currently outnumber Painted Turtles 9:1 on average among these sites (A. Mitchell, Coastal Painted Turtle Project, Vancouver, BC, unpublished report 2015). The same situation of predominantly non-native turtles occurs in Seattle, Washington, and Portland, Oregon. Breeding populations of non-native turtles now occur in most of these locations. Confounding matters is the occurrence of other subspecies of Painted Turtles, such as Midland Painted Turtle, *C. p. marginata,* and Eastern Painted Turtle, *C. p. picta,* which have been recorded by researchers in urban settings, due to their relocation and release by people moving from one part of the country to another (Jensen et al. 2014).

Introduced or extralimital species may outcompete or displace native turtles. We lack published evidence of this interaction in the Pacific Northwest; however, there is a growing body of negative impact reported of sliders on

Figure 2. *Left*: Adult Red-eared slider moving overland from a drained stormwater detention pond, Surrey, British Columbia. Photo by Brent M. Matsuda. *Below*: Six shells of Common Snapping Turtles from one lake outside of Portland, Oregon. Photo by Susan Barnes of Oregon Department of Fish and Wildlife.

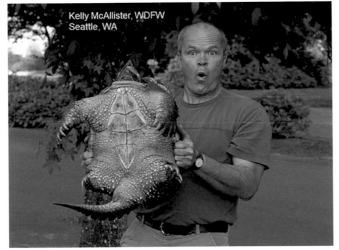

Figure 3. Adult snapping turtle found nesting in backyard in Seattle, Washington. Photo by Kelly McAllister.

native turtles in Europe (e.g., Hidalgo-Vila et al. 2009; Cadi and Joly 2003; Polo-Cavia et al. 2010, 2011) and California (Spinks et al. 2003, Thomson et al. 2010). Some suggested impacts include native turtles avoiding basking sites with sliders, competition for food sources, disease transmission, and fitness. Sliders have higher fecundity than native species (Bury and Germano 2008; Bury et al. 2012) and are known to travel at least 3 miles in search of food, habitat, mates, or nesting sites, often using drainage ditches and creeks, thereby avoiding roads (Dept. of Agriculture and Food 2009). In these cases, any inability to reproduce precludes the title "invasive" (Cadi et al. 2004). Snapping and Softshell turtles can also inflict injury on people who handle them and, because of their secretive nature, populations might become established with little notice.

There is critical need to conduct surveys to better document the presence and abundance of non-native turtles in the Pacific Northwest. Further, we need research to determine how non-native turtles impact native species and trends in population sizes of both groups. Specific studies could determine key population features (e.g., growth rates, fecundity) to better identify their impacts on native turtles. Then, management options may suggest removal or control options for these non-native turtles. Public awareness is key, as removal without education nullifies management efforts (e.g., when turtle populations are continually being replenished by releases from the public sector).

Presence and abundance of non-native turtles in the Pacific Northwest run counter to restrictions on selling them regionally. In the United States, the Food and Drug Administration bans sale of turtles less than 4 inches long, but enforcement may be lax or inconsistent. Regardless of size, Oregon forbids sales of any turtle species that might become established in the state. Still, many species of non-native turtles now have breeding populations in these extralimital areas (Spinks et al. 2003, Bettelheim et al. 2006). In BC, sales of all Chelydridae (Common Snapping Turtles), Emydidae ("pond" turtles), and Trionychidae (softshells) have only recently been banned (Wildlife Act 1996, Wildlife Act Designation and Exemption Regulation 1990). Prior to 2012, bans on turtle sales varied by municipal jurisdiction, rendering management ineffective when sliders could simply be purchased in a neighboring area, and enforcement was lax. While it is still legal to possess sliders as pets (Wildlife Act General Regulation 1982), it is illegal to sell, transport, or release them, and non-native turtles can be captured and destroyed without permits. Efforts are needed to better limit inflow of non-native species while removing or controlling those already established. If climate change effects lead to warming temperatures, these conditions favor the fecundity and survival of non-native populations to expand and establish themselves.

Acknowledgments

We thank many individuals for sharing their knowledge, including Kyle Spinks, Tualatin Hills Parks & Recreation District, OR; Susan Barnes and Terry Farrell, Oregon Department of Fish and Wildlife; Kelly McAllister, Washington Department of Fish and Wildlife; Ricardo Small, Albany, OR; and Aimee Mitchell, Coastal Painted Turtle Project, Vancouver, British Columbia.

Turtles (Testudines)

Snapping Turtles: Chelydridae

COMMON SNAPPING TURTLE

Chelydra serpentina (Linnaeus, 1758)

Identification: The Common Snapping Turtle is a typically large, up to 14.0 in., chelydrid turtle. Among adults, males are larger than females. The dorsal background color ranges in shades of brown to black. The tail is long and dorsally saw-toothed. The plastron is reduced, giving the legs great mobility. The toes are webbed, and the claws are long and sharp. Combined with a long neck, very strong jaws, and fast reflexes, a large and frightened Common Snapping Turtle can be a potentially dangerous undertaking if one is not careful in handling. Handling an individual by its tail can be harmful to the turtle. Two forms of the Common Snapping Turtle exist: the Northern Snapping Turtle, *C. s. serpentina*, and the Florida Snapping Turtle, *C. s. osceola*. The Common Snapping Turtle is an eastern North American species.

Introduction history and geographic range: In the United States, the Common Snapping Turtle is exotic to Arizona, California, Nevada, New Mexico, Oregon, Utah, and Washington. A potentially abundant source of food for human

Photo courtesy of Eugene Wingert.

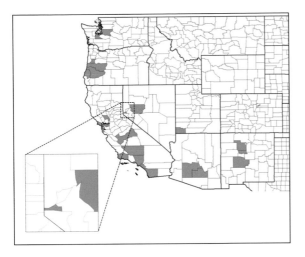

consumption, this species has been transported outside its native range. Creation of artificial ponds has increased the rate of its dispersal.

Ecology: The Common Snapping Turtle can inhabit most any kind of freshwater and brackish habitats, but it most prefers still water less than 36.0 in. deep, with lots of underwater structure. Northern populations go dormant in winter, and mating can take place anytime during the active season. June–July nesting is typical of the species but is longer in southern latitudes. One large clutch of 25–40 eggs each year is typical, and the round, hard-shelled eggs hatch about three months later. Maturity is reached in 4–13 years, depending on latitude, and at 8.0–10.0 in. The Common Snapping turtle can be the top predator in many aquatic systems, where large prey are drowned and rendered. Much of what is eaten is often invertebrates and fish. Seasonally, calling anurans are hunted. Eggs can be subject to intense predation. Young turtles are at risk of many predators, including conspecifics. Once an adult, this turtle's primary source of mortality is humans. Common Snapping Turtles will move extensive overland distances between aquatic habitats. If even for a short time, its visit to an otherwise fishless wetland can expose native aquatic species to predation by an evolutionary stranger.

Box and Water Turtles: Emydidae

SOUTHERN PAINTED TURTLE

Chrysemys dorsalis Agassiz, 1857

Identification: The Southern Painted Turtle is a small, up to 5.0 in., emydid turtle. Among adults, males are smaller than females and can be identified by a long, thick tail and long foreclaws, which are used in courtship. The dorsal

Photo by Suzanne Collins.

background color ranges from brown to black, and an orange to red stripe runs along the mid-dorsal aspect of the carapace. The edges of the pleural scutes are thick and olive in color. The plastron is uniformly yellow. The Southern Painted Turtle is the southernmost form of a closely related group that includes the Western Painted Turtle, *C. picta belli,* and Midland Painted Turtle, *C. p. marginata.* All the forms are easily identifiable; however, hybridization and intergradation occur, which obscure distinction in pattern. The Southern Painted Turtle is native to portions of Alabama west to eastern Texas and north from coastal Louisiana to southern Illinois in the United States.

Introduction history and geographic range: In the United States, the Southern Painted Turtle is established in a pond in southern Florida (Miami-Dade County) and a record exists for Alachua County.

Ecology: The Southern Painted Turtle inhabits soft-bottomed lentic or sluggish lotic freshwater systems. Northern populations emerge in the spring later than those farther south. Individuals readily bask and often in large groups. An average of approximately eight eggs are deposited in each clutch. One or two clutches are typically produced annually and hatch in about three months. Painted Turtles mature in three to six years and can live 20–30 years in the wild, with some possibly living to 60 years. An omnivore, its diet includes vegetation and small aquatic vertebrates and invertebrates.

WESTERN PAINTED TURTLE
Chrysemys picta belli (Gray, 1831)

Identification: The Western Painted Turtle is a large, up to 8.0 in., emydid turtle. Among adults, males are smaller than females, and males can be identified by a long, thick tail and long foreclaws, which are used in courtship. The dorsal background color ranges in shades of light brown to olive. Bars are present along the marginal scutes, and the edge of the carapace is ringed in light red. The plastron is light with irregular dark shapes connecting to seams. The Western Painted Turtle is the westernmost form of the polytypic group of the Painted Turtle, *C. picta,* that includes the Midland Painted Turtle, *C. p. marginata,* and the Eastern Painted Turtle, *C. p. picta,* and a close relative, the Southern Painted Turtle, *C. dorsalis.* All are easily identifiable; however, hybridization and intergradation occur, which obscures distinction in pattern. The Painted Turtle is native primarily to the north-central portion of the United States.

Photo by Suzanne Collins.

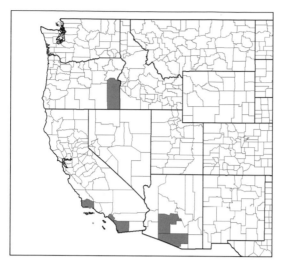

Introduction history and geographic range: In the United States, the Western Painted Turtle is exotic to Arizona (Maricopa and Pima counties), California (Orange, San Diego, and Santa Barbara counties), and Oregon (Malheur County).

Ecology: The Western Painted Turtle inhabits soft-bottomed lentic or sluggish lotic freshwater systems. Northern populations emerge in the spring later than those farther south. Individuals readily bask and often in large groups. An average of approximately 12 eggs are deposited in each clutch, which is larger than averages produced by other species of Painted Turtles. One or two clutches are typically produced annually in the North and hatch in about 2.5 months. Painted Turtles mature in three to six years and can live 20–30 years in the wild, with some possibly living to 60 years. An omnivore, its diet includes vegetation and small aquatic vertebrates and invertebrates.

FALSE MAP TURTLE
Graptemys pseudogeographica (Gray, 1831)

Identification: The False Map Turtle is a large, up to 10.0 in., emydid turtle. Females are substantially larger than males. Among adults, males can be identified by a long, thick tail and long foreclaws, which are used in courtship. The carapace is keeled, or "saw-backed" and brown in color, with yellow to white markings. Both the low black spines on the carapace and the vibrant pattern on plastron fade as turtles reach adulthood. The skin color is olive with light striations, and the neck is striped. The False Map Turtle, *G. pseudogeographica*, comprises two regionally distinct forms, the Northern False Map Turtle, *G. p. pseudogeographica,* and the Mississippi Map Turtle, *G. p. kohnii.* The character

Photo by Suzanne Collins.

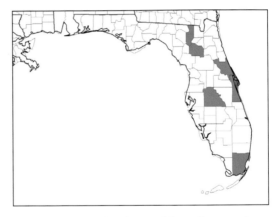

that most easily distinguishes the two forms is the shape of the yellow marking behind the eye. A distinct yellow spot is present behind the eye of the nominate form, whereas the Mississippi Map Turtle has a well-defined crescent that encircles the posterior portion of the eye and interrupts the neck stripes. The False Map Turtle is native to the central United States.

Introduction history and geographic range: In the United States, the False Map Turtle is exotic to a few ponds on the Florida International University (FIU) campus in Miami, Miami-Dade County, Florida. This colony was the result of introductions or escapes of individuals from ponds on the FIU Modesto Madique Campus. Records exist for five other counties.

Ecology: Strictly a freshwater turtle, this strong swimmer most prefers slow-moving waters of big creeks and rivers, although it may be found in some lentic situations. Aquatic vegetation and basking sites well away from shore are important to this species. Up to nine hours off and on during each day are spent basking, often very close to one another and in large numbers. Northern populations go dormant in winter. In Wisconsin, egg-laying occurs from May through July. Two and maybe three clutches can be produced annually. Clutch sizes of the Northern False Map Turtle average 14.1 eggs, with a range of 8–22 eggs. The Mississippi Map Turtle produces clutches ranging from 2 to 8 eggs. Depending on temperature inside the nest chamber, incubation can average 52.1–89.3 days. In Wisconsin, sexual maturity is reached in six years in males and eight years in females. Both subspecies are omnivorous. The Northern False Map Turtle feeds on vegetation and aquatic invertebrates, and, depending on location, it may specialize in eating mollusks. The increasing popularity of this species in the pet trade has resulted in numerous observations of this turtle elsewhere well outside its natural range.

FLORIDA RED-BELLIED COOTER
Pseudemys nelsoni Carr, 1938

Identification: The Florida Red-bellied Cooter is a large, up to 12.0 in., emydid turtle. Among adults, males are smaller than females. Males have a longer tail than females and have long foreclaws that are used in courtship. The Florida Red-bellied Cooter is thick-shelled with a somewhat higher domed carapace than other cooters. The background color of the carapace is dark brown interrupted by a vertical line, rusty-red in color, on each pleural scute. The carapace is often at least partially covered in algae. The bridge of the shell is lined with large solid black spots. The plastron is unmarked, and both the plastron and the bridge are a light to dark orange or orange-red. The jaw is notched. The carapace of hatchlings is heavily patterned in green and black; the plastron is vibrant red

Photo courtesy of Gary Busch.

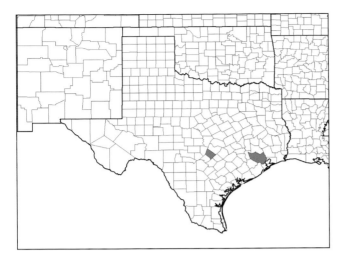

and intermittently blotched in black along the sutures. The Florida Red-bellied Cooter is native to Florida and extreme southeastern Georgia.

Introduction history and geographic range: In the United States, the Florida Red-bellied Cooter is exotic to Aquarena Springs in San Marcos, Hays County, Texas, where it was reported in 1998 and subsequently in Harris County.

Ecology: The Florida Red-bellied Cooter is strictly a freshwater turtle and a regular basker in preferably shallow, well-vegetated wetlands. Its thick shell and domed carapace provide some protection from predation by the American Alligator in the relatively shallow weedy systems it prefers. Eggs are deposited in the summer, often in nests of the American Alligator. Incubation time is typically three months in duration. Hatchlings measure about 1.25 in. Juveniles are omnivorous and shift to a diet heavily comprising vegetation when adult. Predation is high on eggs, and young individuals are subject to predation by a wide range of predators.

Northern Red-bellied Cooter
Pseudemys rubriventris (LeConte, 1830)

Identification: The Northern Red-bellied Cooter is a large, up to 14.0 in., emydid turtle. Among adults, males are slightly smaller than females. Males have a longer tail than females and long foreclaws, which are used in courtship. The background color of the carapace is dark brown or black and is interrupted by a vertical line, red in color, on each pleural scute. Marginals have a red bar and a dark blotch with a light center below. Individuals often darken with age. The bridge is crossed with a long dark bar. The plastron is unmarked, and both the plastron and the bridge are orange to salmon and may have faded black markings along

Photo courtesy of Eugene Wingert.

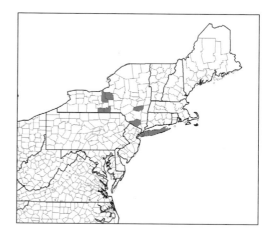

the seams. The jaw is notched. The carapace of hatchlings is heavily patterned in green and black; the plastron is vibrant red and intermittently blotched in black along the sutures. The Northern Red-bellied Cooter is native to the Atlantic Coastal Plain from northeastern North Carolina through Virginia and New Jersey and up the Potomac to West Virginia. A disjunct population occurs in Massachusetts.

Introduction history and geographic range: In the United States, the Northern Red-bellied Cooter is exotic to 11 counties in New York. On Long Island, it is especially abundant.

Ecology: This turtle occurs in deep, soft-bottomed lentic bodies of freshwater and tidal waters at river mouths. Other important features of its habitat are an abundance of vegetation to eat and basking sites over deep water. In the North, these wary turtles are active from March or April through October. Eggs are deposited from late May through early August, with a peak in June, in the mid-Atlantic states. Clutches average 13 eggs. Eggs hatch from late August to early September. Sexual maturity is reached at 9 years of age in males, and most reproductively active females were at least 11 years old. This species is principally herbivorous. Eggs and young face the most intense predation in a population.

RED-EARED SLIDER
Trachemys scripta elegans (Wied-Newwied, 1838)

Identification: The Red-eared Slider is a large, up to 8.0 in., emydid turtle. Among adults, males average smaller than females. Males have a longer tail than do females and have long foreclaws, which are used in courtship. The carapace is weakly keeled, green in color with light transverse bars. The plastron is marked with black spots of various shapes. Some individuals, especially males, may become melanistic in time, to the extent that it obscures the telltale red "ear" behind the eye. However, this diagnostic feature is not always well defined and occasionally is absent. Young individuals are a more vibrant green than adults. Two forms of the Pond Slider, *T. scripta*, exist: the Yellow-bellied Slider, *T. s. scripta*, and the Red-eared Slider, *T. s. elegans*. The Pond Slider is native to the central states and portions of the Southeast, exclusive of the Appalachian Mountains.

Photo courtesy of Janson Jones.

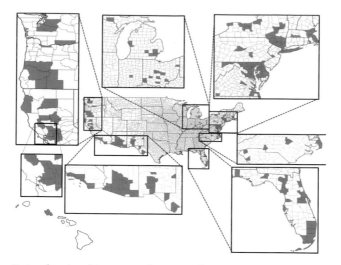

Introduction history and geographic range: In the United States, the Red-eared Slider is exotic to Arizona, California, Delaware, Florida, Hawai'i (Kaua'i, O'ahu), Indiana, Maryland, Massachusetts, Michigan, New Jersey, New Mexico, New York, North Carolina, Ohio, Oregon, Pennsylvania, South Carolina, Texas, Virginia, and Washington. This species has been reported from many other locations and awaits formal confirmation. This highly vagile animal is undoubtedly established in places not yet reported. A popular and inexpensive pet since the early 1900s, this fast-growing, highly fecund, and ecologically generalist turtle continues to be released, often by owners unprepared for its quick growth to a large size.

Ecology: Strictly a freshwater turtle, its favored habitats are still, muddy-bottomed ditches, canals, and ponds having suitable basking logs and much vegetation. Northern populations go dormant in winter. Multiple clutches can be produced in a single season. Depending on location, up to five clutches can be produced, each 12–36 days apart. Six or seven eggs per clutch is typical, but large females can lay up to 23 eggs. Incubation time is typically three months, and some hatchlings may overwinter and emerge the following spring. Maturity is reached between two and five years and earlier in males. Juveniles are carnivorous and eat calcium-rich foods. From a diet comprising primarily insects, the diet shifts with age to primarily vegetation. Adults will eat animal protein when available. Predation can be high on eggs, and young individuals are subject to predation by a wide range of predators. In some parts of the country, basking turtles are routinely shot from bridges for sport. The Red-eared Slider can be pushy when basking, which can place indigenous species at a disadvantage where basking sites are limiting. In places such as Virginia and the panhandle of Florida, extralimital populations of the Red-eared Slider intergrade with the native Yellow-bellied Slider.

Softshells: Trionychidae

FLORIDA SOFTSHELL

Apalone ferox (Schneider, 1783)

Identification: The Florida Softshell is a large-sized, up to 26.5 in., trionychid turtle. Among adults, males are smaller than females. In some places, they are called pancake turtles because of the fleshy, relatively flat carapace. Ridges are present in the nostrils. The carapace is grayish-brown. Faint dark spots may be present. More than one row of spiny-like projections is present on the anterior

Photo courtesy of Janson Jones.

rim of the carapace and extend to the forelegs. The plastron is much reduced in size. The venter is white. Its claws are sharp, its neck long, and jaws strong. Handling even a young individual should be done carefully. Hatchlings are boldly marked with crowded black blotches on an orangish carapace. The carapace is bordered in yellow, and the plastron is gray. A prominent yellow stripe runs from the eye through the side of the neck. The Florida Softshell is native to Florida and portions of South Carolina, Georgia, and Alabama.

Introduction history and geographic range: In the United States, the Florida Softshell is exotic to the Lower Florida Keys in Monroe County and the Rockefeller Wildlife Refuge in Cameron Parish, Louisiana. It was first observed at the southeastern Cameron Parish site in 2001. It was thought to have colonized the canals and impoundments through incidental releases from Florida stock from West Palm Beach, Florida, brought over in the 1970s for culture studies.

Ecology: This highly aquatic species most prefers sandy-bottomed or soft-bottomed shallow lentic bodies of fresh and brackish water. Covered with sand under the water, individuals will periscope with their long necks. Activity is curtailed in the winter in northern areas. Nesting occurs in spring and summer. They deposit an average of 20 eggs, and up to six clutches can be produced annually in Florida. Hatchlings measure approximately 1.5 in. In Florida, diet shifts ontogenetically from insects to mostly snails. Eggs and hatchlings face many predators. Adults are subject to predation by humans and the American Alligator.

SMOOTH SOFTSHELL

Apalone mutica (LeSueur, 1827)

Identification: The Smooth Softshell is a small-sized, up to 14.0 in., trionychid turtle. Among adults, males are smaller than females. In some places, they are called pancake turtles because of the fleshy, relatively flat carapace. Ridges are absent in the nostrils. The carapace is grayish-brown or olive. No spiny-like projections are present on the carapace. Feet are weakly patterned. The plastron is reduced in size. Claws are sharp, neck is long, and jaws are strong. Handling even a young individual is to be done carefully. Hatchlings are marked with slightly darker specks on the carapace. The Smooth Softshell is native to waterways of the Mississippi drainage in much of the central to northeastern United States.

Introduction history and geographic range: In the United States, the Smooth Softshell is exotic to two counties in New Mexico. In Quay County, populations occur in the Canadian River and the Ute Reservoir. In San Miguel County, a population is established in the Conchas Dam.

Photo by Suzanne Collins.

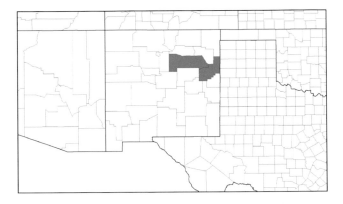

Ecology: This highly aquatic species is closely but not exclusively associated with lotic systems. Activity is curtailed in the winter in northern areas. Nesting occurs in spring and summer. Eggs hatch in late summer approximately three months after oviposition. Its diet comprises mostly aquatic invertebrates, amphibians, and fish; however, fruits and vegetation are consumed occasionally.

Spiny Softshell

Apalone spinifera (LeSueur, 1827)

Identification: The Spiny Softshell is a medium to large-sized, up to 21.0 in., tri-onychid turtle. Among adults, males are smaller than females. In some places, they are called pancake turtles because of the fleshy, relatively flat carapace. Ridges are present in the nostrils. The light brown or olive color of the carapace is covered with dark ocelli in juveniles and males and is camouflaged in dark blotches in females. The carapace of the male has the texture of sandpaper. In both sexes of most populations, one row of spiny projections is present on the

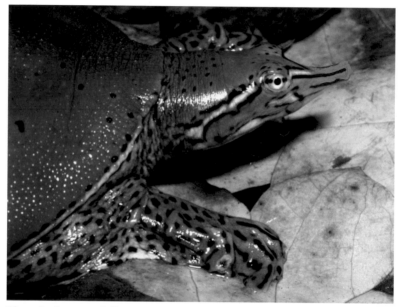

Photo by Suzanne Collins.

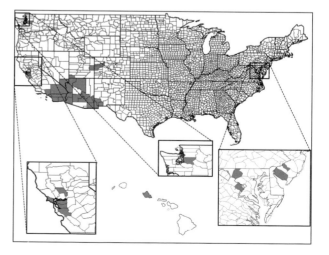

anterior rim of the carapace. Its venter is white. The plastron is much reduced in size. Its claws are sharp, its neck long, and its jaws strong. Handling even a young individual should be done carefully. Three forms of this polytypic species are recognized: the Eastern Spiny Softshell, *A. s. spinifera,* the Gulf Coast Spiny Softshell, *A. s. aspera,* and the Texas Spiny Softshell, *A. s. emoryi.* The Spiny Softshell is native to much of the central and portions of the eastern and southwestern United States. The Spiny Softshell occurs naturally in the Maurice River system, Raritan River watershed, Cooper's Creek, and Palatine and Rainbow Lakes of New Jersey, and in ponds in Virginia.

Introduction history and geographic range: In the United States, it is exotic to Arizona, California, Colorado (western), Hawai'i (O'ahu), Nevada, New Jersey, New Mexico, Utah, Virginia, and Washington. California populations comprise the Texas Spiny Softshell. Much of the western extralimital range of the Spiny Softshell is associated with the Gila-Colorado River system. Its introduction into Colorado River systems was derived from animals of the Gila River Drainage of Arizona and New Mexico in 1900. In California, it is found in the San Gabriel and San Diego rivers, as well as in ponds and reservoirs, such as the Lower Otay and San Pablo reservoirs.

Ecology: This highly aquatic species mostly prefers rivers and other lotic bodies of primarily, but not exclusively, freshwater. In the East, this species will use ponds as well. Northern populations go dormant in winter. Nesting occurs during May–August, with most activity during June–July. Up to 39 eggs have been reported from a single clutch; however, 12–18 eggs is typical for the species. Two clutches are produced annually. Incubation is approximately 90 days, and in some populations, hatchlings may overwinter and emerge the following spring. Its diet is composed of aquatic vertebrates and invertebrates, especially crayfish. Eggs and hatchlings face many predators. Adults are subject to predation by humans and the American Alligator. Effects of this turtle on indigenous populations of crayfish and crayfish predators warrant study.

WATTLE-NECKED SOFTSHELL
Palea steindachneri (Siebenrock, 1906)

Identification: The Wattle-necked Softshell is a small to medium-sized, up to 16.0 in., trionychid turtle. Among adults, males are smaller than females. Well-developed tubercles are clustered where the carapace meets the neck, hence the name. The carapace is uniformly olive brown in adults and speckled black in juveniles. Among adults, the venter is pink and gray, whereas that of juveniles is yellow in color. The plastron is much reduced for greater mobility. Its claws are sharp, its neck long, and its jaws strong. A bite by an adult can be very painful. Its center of distribution is southeastern China and northern Vietnam.

Photo courtesy of Torsten Blanck.

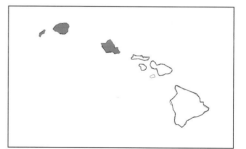

Introduction history and geographic range: In the United States, the Wattle-necked Softshell is exotic to Hawai'i (Kaua'i and O'ahu), where individuals were brought for human consumption from at least the early 1900s up until the advent of the Second World War.

Ecology: In Hawai'i, this highly aquatic turtle can be found in a wide range of systems, including ponds, marshes, canals, and streams. Up to 28 eggs are deposited near the water line. Up to four clutches can be produced each year, and the eggs require up to 68 days to hatch. Maturity is reached at four or five years of age. The Wattle-necked Softshell is a carnivore, readily consuming fish and aquatic invertebrates. Both the North American Bullfrog and South American Cane Toad are reported prey of this species in Hawai'i. In Hawai'i, the larger-bodied Wattle-necked Softshell appears to be the more successful of the two introduced species of Asian softshells. The Wattle-necked Softshell is very popular for human consumption and it is very fecund. Ecologically, it is easy to please, and no suitable rationale exists to risk colonization on the mainland through its importation.

CHINESE SOFTSHELL

Pelodiscus sinensis (Wiegmann, 1835)

Identification: The Chinese Softshell is a small, up to 12.0 in., trionychid turtle. Among adults, males are smaller than females. The drab olive carapace is flat with skin covering bone, except along the edges. It may or may not have darker blotches on it. Light-bordered black spots are present on the carapaces of juveniles. The venter is whitish. The plastron is much reduced for greater mobility. Its center of distribution is southern and central China, Hainan and Taiwan, and Vietnam.

Photo courtesy of Tien-His Chen.

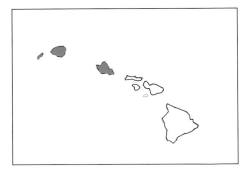

Introduction history and geographic range: In the United States, the Chinese Softshell is found in Hawai'i (Kaua'i and O'ahu), where it was introduced as a source of meat for human consumption from the late 1800s up until the outbreak of the Second World War.

Ecology: The Chinese Softshell is quite an aquatic generalist in its habitat requirements. It can be found in both slow-moving and still water in Hawai'i. This is a highly fecund turtle. Up to 30 eggs can be produced in a single clutch, and several clutches can be produced each year. Eggs hatch in about 60 days. The diet comprises primarily aquatic invertebrates, although it may actively eat plants as well. In Hawai'i, the Chinese Softshell appears to be less successful than the larger-bodied Wattle-necked Softshell.

PART 4

LIZARDS

Preceding the following species accounts, we present three invited essays. The first, by Jesse Rothacker, discusses trends and concerns relating to ownership of exotic reptiles from the standpoint of a rescue operation in south-central Pennsylvania, one of many such outfits in the United States that play an important role in reducing and preventing the release of non-native animals by offering to provide a refuge for them as an alternative. The second essay, by Frank J. Mazzotti, using the introduction and establishment of the Argentine Giant Tegu in Florida as an example, illustrates the easily preventable delays that hinder effective action in addressing invasive species. The third essay, by Walter E. Meshaka Jr., discusses the value of standardized counts of animals and applies that approach to an urban herpetofauna dominated by exotic species.

Observations from a Reptile Sanctuary in South-Central Pennsylvania

JESSE ROTHACKER

Since 2004 I have run a nonprofit reptile rescue and educational organization called Forgotten Friend Reptile Sanctuary (FFS). A big part of our work is finding new homes for unwanted amphibians and reptiles from the pet trade. Although there are hundreds of exotic herp species in the pet trade, there are fewer than a dozen species that fill up rescues. Our 17 years of experience have seen the Red-eared Slider, *Trachymys scripta elegans,* as the most unwanted species, followed by the Green Iguana, *Iguana iguana.* To that end, during 2013–2014 combined, intake requests numbered 159 for Red-eared Sliders and 50 for Green Iguanas. Some of the next most common are Red-tailed Boa Constrictors, *Boa constrictor constrictor* and *B. c. imperator,* Ball Pythons, *Python regius,* Central Bearded Dragons, *Pogona vitticeps,* Leopard Geckos, *Eublepharis macularius,* American Alligators, *Alligator mississippiensis,* African Spurred Tortoises, *Centrochelys sulcata,* and Russian Tortoises, *Agrionemys horsfieldii.* With less frequency, we also see Carpet Pythons, *Morelia spilota,* Burmese Pythons, *Python bivittatus,* Kingsnakes, *Lampropeltis* species, Red Cornsnakes, *Pantherophis guttatus,* and a few other odd species.

A few factors seem to be associated with certain species being discarded much more so than other species. Most often, the species that are inexpensive to bring home but expensive to keep are the most likely to soon outgrow their welcome. To provide but one example, the Red-eared Slider may be purchased for as little as one dollar, or certainly for less than 20 dollars. A complete setup for a pet turtle, however, includes many expensive parts. There is the cost of a large tank or backyard pond, special lighting, special pumps and filters, basking areas, and other habitat enrichments. Owners who bring home this inexpensive baby turtle soon find out to their dismay that their pet demands an investment of supplies of several hundred dollars or more, and the turtle may outlive them. Furthermore, these turtles live in their own toilets. Without an expensive filter and regular water changes, the tanks will quickly become effectively unflushed and very unsanitary toilets.

Although hardly anyone would bring home a puppy from the pet store without serious long-term consideration, many people think of a pet reptile as low maintenance without need of daily attention. Before long, new pet owners, not having done their homework, find out that in some ways captive reptiles are even more demanding in time and money than their canine or feline counterparts.

Long-term financial costs are the top reason that reptiles are brought to FFS, and large adult sizes and long life spans contribute to these costs. Boa Constrictors, American Alligators, and Green Iguanas are great examples of reptiles being discarded because of unexpected adult body sizes. All three of these species fit in the palm of your hand as babies. Many pet owners literally bring these pet reptiles home from the store inside their pockets, because baby reptiles are so tiny. However, each of these three species grows to very large adult sizes. An owner may bring home a new baby reptile with optimistic visions of building a giant enclosure, perhaps years down the road. But in many cases, the animal becomes surprisingly more challenging after a few years of rapid growth, and the optimistic visions disappear. We tend to see most large species come into our rescue once they leave their "baby size" and are subadults. It is common for us to have Boa Constrictors, which are born around 10 inches long, surrendered to us at 3–5 feet in total length. At this point they are no longer eating baby mice, and owners are struggling to find full-grown rats to feed them and imagining how they're going to find even bigger food items in a few years, like rabbits or chickens. Likewise, Green Iguanas, which start around the size of a mouse, most often arrive to us at 3–4 feet in total length. This body size is the threshold beyond which the iguana has outgrown its first tank or cage enclosure and is now ready for its own bedroom. Most alligators are purchased as babies, around 5–6 inches long. We tend to receive them between the lengths of 1 and 3 feet. Increasing body size can be associated with behaviors that place the keeper at risk. A 6-ft. Boa Constrictor should not be held while alone. A defensive 4–6-ft. Green Iguana is actually dangerous with claws, teeth, and tail. A Burmese Python can reach 6 feet in total length in one year, and an American Alligator can reach 4 feet in a few years, to the surprise of the unprepared keeper. While most cats and dogs live 10–15 years, many reptiles may live several decades, some beyond 100 years. We have received many turtles and tortoises from senior citizens who brought home their pets 20, 30, or even 40 years earlier!

In summary, the main reasons we see specific reptiles given up to our rescue are because they are inexpensive to bring home, expensive to keep, grow quickly to undesirable sizes, and they may live several times longer than other pets. When an animal is treated as a disposable object with little more than impulse behind its purchase, we, or the natural system around us, will be the recipient of it soon enough. Rare is the rescue animal from a person who has done their due diligence in the care and long-term cost of keeping their pet.

The Story of the Argentine Giant Tegu, *Salvator merianae,* in South Florida

When a New Invader Meets Bureaucratic Reality, Who Will Win?

FRANK J. MAZZOTTI

Florida has more non-native reptiles and amphibians than anywhere else in the world, with at least 180 introduced species, more than 60 of which are established species (i.e., locally breeding). Florida has a greater diversity of established non-native lizards (44 species) than native lizards (40 species), and the largest lizards are all non-native species. The potential for one of these species to turn into the next Burmese Python concerning detrimental impacts is real. Why is Florida so vulnerable to invasion; why do some species seem preadapted to be successful invaders (spreading and causing ecological, economic, or social harm); and based on our experience with Burmese Pythons are we now prepared to repel a new invader?

South Florida has proven particularly vulnerable to invasion by reptiles because it has a peninsular geography, tropical climate, a disturbed natural environment, and major sources of non-native species from the pet trade (port of entry, captive breeders, and animal dealers). Although it is clear that the current source of invasive reptiles is the pet trade, it is not clear which introductions have been the result of deliberate releases or unintentional escapes.

It is well known that many reptiles, especially snakes and lizards, are escape artists and require sturdy, secure enclosures to contain them. This is particularly important in Florida, where a tropical climate allows pet reptiles to be kept outside and provides occasional hurricanes that increase the risk of escape. Deliberate releases seem to have multiple motivations.

Reptiles that get large, particularly those that are carnivorous, can become difficult to maintain in captivity. They require large spaces and lots of food, and they can produce prodigious amounts of excrement. A combination of being unable to continue to maintain a pet and unwilling to euthanize it has led some owners to release their pets into what they perceive as a warm welcoming environment in South Florida. Education and the establishment of "Exotic Pet Amnesty Days," when non-native pets can be surrendered by owners without fault, by the Florida Fish and Wildlife Conservation Commission (FWC) have provided an alternative to releasing pets.

Pet dealers release specimens that are damaged and unsellable, sometimes with the intention of starting a breeding colony that could later be harvested for sale. Individuals will also start colonies for the purpose of future harvest, or increasingly for aesthetic purposes; that is, people just want to be able to go see an exotic species. In the case of Argentine Giant Tegus, *Salvator merianae,* in extreme South Florida, the tegus apparently escaped from a facility where they were kept negligently in open cages and allowed to run freely.

Non-native reptiles that are released or escape are met with a benign, welcoming, tropical climate in South Florida. There is a distinct May to October rainy season. During the winter, daytime temperatures average around 77°F and nighttime temperatures average around 63°F. Importantly for many non-native reptiles is that temperatures rarely, only a few times a decade, go below 45°F. Temperatures around 41°F cause green iguanas to go into torpor and fall out of trees, and freezing temperatures have killed Burmese Pythons.

South Florida has a depauperate (relatively few species) herpetofauna (amphibians and reptiles) as a result of being at the tip of a peninsula. Dominated by Greater Everglades ecosystems, South Florida offers a combination of human-dominated and disturbed natural ecosystems that provide open niches for invading reptiles. Hence non-native wildlife invading South Florida is met with warm temperatures, plenty of water, few competitors, and lots of available housing. Is it any wonder that so many non-native species successfully establish in South Florida?

The characteristics of invasive reptiles have changed over the past 30 years. Initially, invasive reptiles were small, early maturing, insectivorous lizards known to be associated with human dwellings, such as Cuban Brown Anoles, *Anolis sagrei,* and Mediterranean Geckos, *Hemidactylus turcicus.* Most were introduced as a result of cargo trade. Today, successful invaders are large bodied, early maturing, prolific, omnivorous lizards that thrive in a variety of habitats. Most were introduced as a result of the pet trade. Argentine Giant Tegus are a perfect example of the new invader.

Argentine Giant Tegus are a large (3–4 ft.) teiid lizard widely distributed in central South America, including warm temperate areas. Argentine Giant Tegus are hunted for hides and meat in South America and have been well established in the pet trade in the United States. As a result of the pet trade, Argentine Giant Tegus have established disjunct breeding populations in parts of southern Miami-Dade, Hillsborough, Charlotte, and Polk counties in Florida. They are primarily terrestrial, although Argentine Giant Tegus have been observed associated with aquatic systems. Argentine Giant Tegus are habitat generalists and are observed in coastal areas, clearings, edges, and frequently in disturbed areas. These lizards are active burrowers and spend substantial time in burrows, including extended winter brumation periods. Argentine Giant Tegus exhibit an omnivorous diet, including vegetation, fruits, seeds, snails, arthropods, fish, birds and bird eggs, small mammals, amphibians, reptiles and reptile eggs, and carrion. In addition to being a large, fecund, omnivorous, habitat generalist of a lizard, the Argentine Giant Tegu is vagile and known to have become successfully established on the Fernando de Noronha Archipelago in Brazil after a deliberate introduction to control rodent populations. Together, these traits make Argentine Giant Tegu a formidable invader.

The South Florida population is located in extreme southern Miami-Dade County. The epicenter is south of Florida City. Argentine Giant Tegus in South Florida are of particular concern because of their proximity to Everglades National Park (ENP), Biscayne National Park, American Crocodile (*Crocodylus acutus*) nests at the Florida Power and Light Company Turkey Point Power Plant site, and several Everglades restoration projects. The population was likely introduced/established around 20 years ago and by 2008 had come to the attention of partners of the Everglades Cooperative Invasive Species Management Area (ECISMA). The invasion of Burmese Pythons had caught us by surprise. No one was prepared for the establishment, proliferation, spread, and degree of impact that followed the introduction of Burmese Pythons into Everglades environments. Surprise combined with the extremely cryptic nature of pythons and the lack of available tools and resources put us at a distinct disadvantage in fighting that invasive species. Argentine Giant Tegus offered us the opportunity to demonstrate that we had learned from our experience, and that we would vigorously amass our resources and contain if not eradicate this new invader.

Interim reviews on our tegu response are mixed. Overall, our response started slow and so far can be best described as too little, too late. That said, at every step in this invasion, all the agencies and individuals involved have done everything they could with the resources available to them to combat the tegu invasion. The problem has been lack of resources necessary to effectively deal with tegus. That can still change.

Figure 4. Points of interest in the South Florida range of the introduced Argentine Giant Tegu, *Salvator merianae*. Photo by Frank J. Mazzotti.

The response to sightings of Argentine Giant Tegus started slowly because the primary responders had day jobs, but a sense of concern and responsibility prompted action. Starting in 2009, ECISMA partners scraped together available resources and began surveillance of tegus using camera traps and by 2010 were live trapping in areas where photographs of tegus had been obtained. This led to an effort to radio-track a female tegu, which resulted in the discovery of the first known nest in South Florida in 2011. In 2012 a cooperative trapping program was initiated by the FWC, and in 2013 at least two Argentine Giant Tegus were photographed emerging from an American Alligator (*Alligator mississippiensis*) nest, each lizard with an egg in its mouth.

In 2014 trapping started to get more systematic in the Southern Glades Wildlife Environmental Area and along the boundary with Everglades National Park. In 2016 trapping was expanded eastward toward the Turkey Point Power Plant site, with traplines along SR 904 and 137th Ave. managed by the FWC. During this period private trappers were actively trapping on private lands in and around the source area near Florida City, and FWC had a trap loan program that they managed in Florida City and the Redland Agricultural Area (RAA).

During 2012–2019, 5,703 tegus were removed by state, federal, and university partners, while an unknown number were removed by private trappers from private lands during the same period. We do know that one private trapper reported removing several hundred tegus per year for sale in the private pet trade. Several private trappers were working in the area, and the final number of tegus removed by private trappers could have been substantial. Private trapping may have diminished during the past two years as sales of tegus became difficult, and some trappers began to vandalize the traps of others. A recent (2019) increase in tegus trapped on the boundary of ENP has been coincident with the decline in private trapping in the area east of where there has been an increase in tegu captures. In 2020, tegus found their way into ENP, and our efforts to keep tegus from nesting in ENP have failed.

We do not know how tegus made it into ENP, but we can identify gaps in our efforts to contain tegus and hypothesize that those gaps allowed tegus to invade ENP. We can also examine why gaps in the containment effort for tegus occurred. First, what would an ideal response have looked like? The general answer to the question of how to respond to a newly discovered invasive species is rapidly and thoroughly. This response should combine determining the extent (numbers and area) of the invasion, removing the invader as encountered, and post-removal monitoring. If the population of the new invader is low in number and localized in area, eradication may be possible. At some point, depending on the species, number, and extent of invasion, eradication may not be possible and management switches to containment. Although we were prepared in terms of knowing what needed to be done and better equipped to conduct surveillance and removal of tegus than we were of pythons, we were unprepared to amass the necessary funding to do so, a condition that still remains. It is not that we do not know what to do, but that we lack the funding to do it.

There was also slowness in responding, and unfortunately the reasons forthcoming were ones we have heard before. As a result of limited resources, the questions centered on how to know whether tegus will be a successful invader (grow in number, expand area occupied, and have negative impacts). The problem is, and this problem keeps emerging again and again, if we wait until we can answer those questions, it is almost always too late to do something about it. The problem with waiting until we know we have a problem is exacerbated by our tendency to study things prior to action; with invasive species we need to act while we study, or if we have limited resources, we should ask whether a judicious decision resulting in rapid and thorough action might exclude or eliminate an invader. That is, shoot first and ask questions later. When we finally established a serious trapping program in 2014, it was six years after tegus had first been observed. Even then, after being too late, our efforts were too little.

Our containment efforts initially focused on public lands on the southern (Southern Glades Wildlife Environmental Area) and western (ENP) fringes of the source population. Trapping in the RAA was relinquished to an FWC-managed volunteer trap loan program, in which a homeowner would report a tegu and subsequently be loaned a trap and instructed in its use, and to private trappers who concentrated their efforts in a disturbed forested area (a mix of tree farms, groves, and patches of invasive and native plant species) that runs along the southern edge of the RAA from the source population toward and into ENP where it becomes Long Pine Key. We have never quantified the number of tegus removed by private trappers. However, by their own accounts some private trappers decreased their effort in 2018 and 2019 largely because of difficulty in selling tegus coupled with increasing trap vandalism. The increase in tegus trapped along the ENP border was coincident in time and space with a possible decrease in trapping in the adjacent disturbed forested area. We hypothesize that the gap in trapping tegus in the RAA and adjacent disturbed forested area facilitated invasion of ENP.

We have the opportunity to test this hypothesis and to take our last, best chance to prevent expansion of Argentine Giant Tegus. Have we finally learned our lesson, and will we act on our knowledge? What happens in the 2021 tegu season will determine whether we will at least try.

Counting Herps

WALTER E. MESHAKA JR.

Why count herps? That's right, why? For one thing, it's fun and a great way to learn what species are around, when they do various activities, and how they respond to other species and changes in their environment. Herp surveys conducted by riding roads, walking transects or routes, or counting from a boat with a question or two in mind can provide potentially very useful information regarding how these species live and how they respond to changes in their environment. The idea is *not* new, but what you find *is* new, and that can add to ever-increasing awareness of the world around us that we want to better understand.

Case in point: ecological succession. An unmowed lawn can become an old field, then a thicket, and eventually a forest. Each of these stages is an ecological sere. This well-known ecological principle applies uniquely to any habitat, including urban human habitats. In the early 2000s, a residential development was constructed in Miramar, Florida. Well, starting with "I wonder what's here?" I got curious to know what herps inhabited the development and began a standardized walk at different times of day and in different months shortly after its construction. Very quickly, I found that most of the herp species were exotic and generally numerous. The opposite applied to native species. As I gave it some thought, it made sense. The development was not constructed on previously intact habitat; rather, it was constructed on habitat that had been disturbed for a long time. The habitat was marginal for native species I could detect and had a robust reservoir of exotic species. Consequently, unlike construction of old, this residential community represented what is typical now, whereby it was constructed with exotic species already there and facing diminished native predators and competitors. The difference was apparent even in some of the exotic species. For instance, not so long ago the Mediterranean Gecko, *Hemidactylus turcicus,* or, more often,

Figure 5. Residential study site in Miramar, Broward County, Florida, photographed on 15 October 2017, after Hurricane Irma passed through on 10 September. Photograph by W. E. Meshaka Jr.

Figure 6. Residential study site in Miramar, Broward County, Florida, photographed on 2 November 2017, after Hurricane Irma passed through on 10 September. Photograph by W. E. Meshaka Jr.

the Indo-Pacific Gecko, *H. garnotii,* would appear on buildings new or old. Then the Wood Slave, *H. mabouia,* would appear and replace both of them, often in big numbers. The Miramar site, on the other hand, began with only Wood Slaves, and lots of them.

Well, I got curious. How do the various species sort out with respect to time of day, temperature, time of the year, and perch height? How well will the various species fare over time? I asked some of these questions of some of their avian predators and feral cats. A few times each year beginning in 2006 I began recording data on them and could spend many days at different times walking my 2-mile route, sometimes between hurricanes (figures 5–6). So far, my last visit was in 2020 and I filled my Excel rows and columns with data to answer my questions. Some questions I just picked up along the way; some questions I've published along the way. The Wood Slave remains the only gecko established there. I am not surprised, but this finding corroborates other findings of a taxon cycle with hemidactyline geckos in Florida. Listening for frog calls at night and walking along shorelines, I've learned a lot more about the breeding season of the Cane Toad, *Rhinella marina* (figures 7 and 8) and the weather conditions associated with it.

Figure 7. A breeding site for the Cane Toad at a residential study site in Miramar, Broward County, Florida, photographed 17 October 2018. Photograph by W. E. Meshaka Jr.

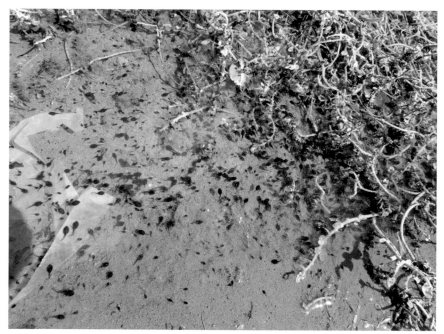

Figure 8. Multiple size-classes of tadpoles of the Cane Toad, *Rhinella marina*, at a residential study site in Miramar, Broward County, Florida, photographed on 24 March 2018. Photograph by W. E. Meshaka Jr.

Figure 9. Close-up of a male North American Green Anole, *Anolis carolinensis*, at a residential study site in Miramar, Broward County, Florida, photographed on 7 September 2019. Photograph by W. E. Meshaka Jr.

Some species have come and gone. Early on, a few species, like the Giant Ameiva, *Ameiva ameiva,* during January–December 2008, and the Bark Anole, *Anolis distichus,* one individual seen once in 2008, made brief appearances, not to be seen again. The Northern Curly-tailed Lizard, *Leiocephalus carinatus,* likewise appeared briefly and disappeared, one individual seen once in 2010, but then reappeared much later along my route in low but steady numbers during 2017–2020, and quickly colonized other areas of the neighborhood and in high numbers. That of course, led to another herp counting project, and I'm comparing their ecology to that of the Brown Anole, *A. sagrei,* that for now at least co-occurs with the Northern Curly-tailed Lizard in these sunnier portions of the development. It will be interesting to see whether the presence of this new predator results in changes to aspects of the anole's behavior or even to its population size and structure compared with what I see along my route.

The Everglades Racer, *Coluber constrictor paludicola,* persists ironically in its destroyed namesake as do the North American Green Anole, *A. carolinensis* (Figure 9), and the Florida Cottonmouth, *Agkistrodon conanti.* Among the aquatic turtles, the Florida Red-bellied Cooter, *Pseudemys nelsoni,* and the Peninsula Cooter, *P. peninsularis,* persist. I wish there were more there to learn more about. Still, such remain the seeds of potential colonizations by these species under the right conditions. What ecological sere would that be and for how long? Time will tell along with counting.

Lizards (Squamata)

Agamid Lizards: Agamidae

Peter's Rock Agama

Agama picticauda Peters, 1877

Identification: Peter's Rock Agama is a medium-sized, up to 12.0 in., agamid lizard. Among adults, males are larger than females. Adult males are blue-black in color with a yellow head and nape and yellowish tail. Adult females and juveniles are brown overall. Its center of distribution is East Africa.

Introduction history and geographic range: Populations of this species in the United States were derived from the pet trade and have been established in southern Florida since the mid-1980s, where it was reported from two sites in Davie, Broward County, Florida, in 1999. Subsequent dispersal through intentional release is common and often in disparate locations. In the United States, Peter's Rock Agama is found in a somewhat scattershot pattern in Florida.

Ecology: Peter's Rock Agama is semi-arboreal and can be found on the ground and on rocks, trees, and buildings. This species is diurnal in activity. It is

Photo by Suzanne Collins.

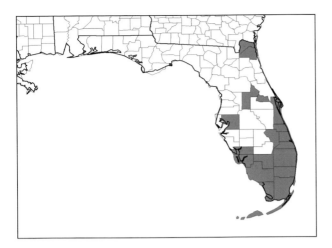

unknown in natural systems of Florida. Population sizes can be large. Although wary of human approach, individuals perch prominently in the open, especially in hot conditions. In southern Florida, females are gravid during May–August and can produce an average of nine eggs up to three times each year. Peter's Rock Agama eats insects and small vertebrates. Potentially as a result of predation, the Brown Anole is uncommon in otherwise acceptable habitat colonized by Peter's Rock Agama. If this is so, other anoles, including the native North American Green Anole, are at risk of predation.

INDO-CHINESE BLOODSUCKER

Calotes mystaceus Duméril and Bibron, 1837

Identification: The Indo-Chinese Bloodsucker is a medium-sized, up to 15.0 in., agamid lizard. Among adults, males are larger than females. This species belongs to a taxonomically complicated group. Bloodsuckers have a distinctive dorsal crest of spines. Males are gray with blue heads. During the breeding season, the male's throat may turn orange. Females are brownish with dark crossbars. Its center of distribution is southeastern Asia.

Introduction history and geographic range: Populations of the Indo-Chinese Bloodsucker in the United States were derived from the pet trade and have been established in southern Florida since the early 1980s, where it was reported from Okeechobee, Okeechobee County, in 1999. In the United States, the Indo-Chinese Bloodsucker is localized in southern Florida.

Ecology: The Indo-Chinese Bloodsucker is terrestrial-arboreal in habits, although it much prefers large trees. Naturally occurring in forests, this species can adapt easily to tree-lined streets, parks, and such. In Florida, it is associated primarily with a specific citrus grove. It is strictly diurnal and is an insectivore.

"Calotes mystaceus manipur" by Ianaré Sévi is licensed under CC BY 3.0. https://commons.wikimedia.org/wiki/File:Calotes_mystaceus_manipur.JPG.

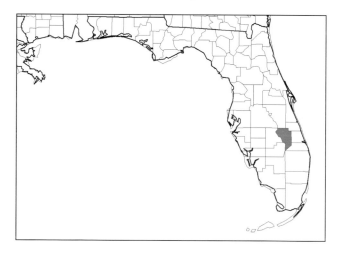

During the breeding season, territorial males perform showy displays. Females lay shelled eggs. Little else is known about the ecology of the introduced population. Ecologically, it might best be considered analogous to the anoles. Impacts of this predator are unknown. This species has yet to prove itself comparatively as successful a colonizing species as its congener, the Variable Bloodsucker, and the reasons for its limited geographic distribution are unknown. For these reasons, perhaps with perseverance this species can be extirpated from Florida.

VARIABLE BLOODSUCKER

Calotes versicolor Daudin, 1802

Identification: The Variable Bloodsucker is a medium-sized, up to 16.0 in., agamid lizard. Among adults males are larger than females. This species belongs to a taxonomically complicated group. Bloodsuckers have a distinctive dorsal crest of spines. Individuals are indeed variable, and the color can lessen or intensify quickly. Adults are generally drab brown in color with dark blotches along the sides. During the breeding season, the color of the male changes dramatically

"Oriental garden lizard (calotes versicolor) (45114163711).jpg" by Haneesh K M. is marked with CC0 1.0. https://commons.wikimedia.org/w/index.php?curid=76038995.

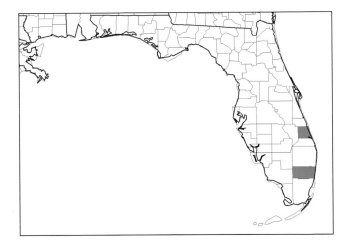

to a yellowish body with a black throat formed by two blotches. Also, the breeding male's shoulders can be crimson or orange. Its center of distribution is Asia.

Introduction history and geographic range: The population of this species in the United States was derived from the pet trade through a shipment from Pakistan in 1978 where it was reported from Port St. Lucie, St. Lucie County, Florida, in 2004. In the United States, the Variable Bloodsucker is established in St. Lucie County, Florida.

Ecology: This species is terrestrial and arboreal in habits. In Florida, it is associated primarily with canal-lined citrus groves and patches of native and exotic vegetation. It is strictly diurnal in activity. During the breeding season, territorial males perform showy displays. Females lay shelled eggs. In southern Florida, a gravid female containing 19 eggs was captured in August as were neonates or young-of-the-year. It is an insectivore and carnivore. Sit-and-wait hunting often takes place in a head-down position on the trunk of a tree. Little else is known about the ecology of the introduced population. Ecologically, it might best be considered analogous to the anoles. Impacts of this predator are unknown, but the versatility in its use of habitat structure and its broad diet increase the likelihood of negative impacts on a wide range of small native and exotic vertebrates through predation and possibly competition for food or habitat. Apparently abundant and readily harvested for the pet trade, the Variable Bloodsucker is the more successful of the two bloodsucker species established in Florida. However, the reasons for its limited geographic distribution are unknown.

BUTTERFLY LIZARD
Leiolepis belliana (Gray, 1827)

Identification: The Butterfly Lizard is a medium-sized, up to 12.0 in., agamid lizard. Among adults, males are larger than females. The lizard is overall flattened in appearance, and among adults, the brown dorsum is covered by small lighter spots that form longitudinal vertebral and paravertebral lines. Orange and black diagonal bands alternate along the sides. Males are brighter in color than females. Hatchlings are striped on a very dark background. Its center of distribution is Asia.

Introduction history and geographic range: The population of this species in the United States was derived from the pet trade, and establishment of this colony, reported in 2005, can be traced to at least 1992 from Miami, Miami-Dade County, Florida. In the United States, the Butterfly Lizard is found in a limited residential area of Miami-Dade County.

Ecology: The Butterfly Lizard is a terrestrial species. It is strongly heliothermic, and individuals prefer open habitat where they very warily bask and actively

Photo courtesy of Kevin M. Enge.

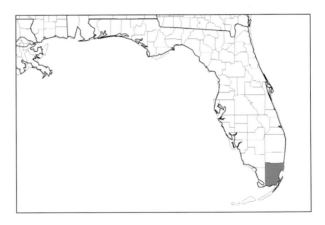

forage in warm conditions. When an individual is frightened, escape is rapid into nearby deep burrows of their own construction. It is strictly diurnal in activity. Females lay shelled eggs. In southern Florida, a gravid female was found to contain five developing ova. It is primarily insectivorous. Little else is known about the ecology of the introduced population. Ecologically, it might best be considered analogous to the Whiptails and *Ameiva*. Impacts of this species are unknown. Individuals are easily captured, and if the introduced range is yet discrete in size, a high likelihood of success exists in the extirpation of this introduced colony.

Chameleons: Chamaeleonidae

VEILED CHAMELEON

Chamaeleo calyptratus Duméril and Bibron, 1851

Identification: The Veiled Chameleon is a large, up to 24.0 in., chamaeleonid lizard. Among adults, males are larger than females. Independently moving eyes, laterally compressed body, a prehensile tail, a tongue that captures and retrieves food, and paired toes are the physical features associated with this species. Adult males are an unmistakable green to turquoise with some dark stippling. Latitudinal orange-yellow bands interrupt a dark lateral band. When in display, a

"Camaleón del Yemen" by Jacinta Lluch Valero is licensed under CC BY-SA 2.0. https://www.flickr.com/photos/70626035@N00/6964597001/in/photolist-bBro3D.

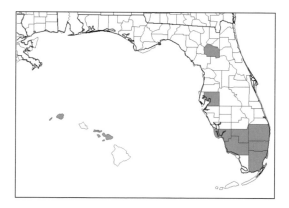

large male is quite striking. The casque on the male can exceed 3.5 in. Tarsal spurs are present on the interior of both rear ankles of males throughout their life. Females are uniform green with changes in spots when gravid and not interested in the male. Their casque is reduced in size. Hatchlings are green with a few creamy white markings. Its center of distribution is the Arabian Peninsula

Introduction history and geographic range: The populations of this species in the United States were derived from the pet trade. In Florida, it was first reported from Ft. Myers, Lee County, in 2004, where it had been in existence since 2001. In Hawai'i, it was first reported from Kaua'i and Maui in 2004, where the species had been present since the 1990s. In the United States, the Veiled Chameleon is found primarily in southern Florida counties and in limited sites in Hawai'i (Kaua'i and Maui).

Ecology: The Veiled Chameleon is a strongly arboreal, diurnal, and cryptic species that has adapted well to the disturbed habitats of its introduced sites, especially Florida. Notwithstanding the effects of a hurricane, the climate of its native range is far harsher with respect to extremes in rainfall and temperature than experienced elsewhere. Captive females can lay several clutches each year numbering 20–50 eggs. Neonates in southwestern Florida have been found in June and August. Captive females can reach maturity in a few months. Although such a rapid growth to maturity is unlikely to be typical in Florida and Hawai'i, Florida colonies are densely populated, and intensive harvesting for the pet trade has been sustainable, indicating rapid recruitment and high fecundity. The Veiled Chameleon is an omnivore. Both invertebrates and vertebrates are readily captured with the use of a protractible tongue. Large prey, such as anoles and grasshoppers, are crushed and swallowed. Individuals also actively seek out flowers and tender leaves. Raptors, corvids, and racers are the likeliest predators of this species. The Florida City site is associated with very few Brown Anoles. No overlap exists with that population and the nearby Oustalet's Chameleon population.

OUSTALET'S CHAMELEON
Furcifer oustaleti (Mocquard, 1894)

Identification: Oustalet's Chameleon is a large, up to 28.0 in., chamaeleonid lizard. Among adults, males are larger than females. Head crest and dorsal spines are more developed in males. Independently moving eyes, laterally compressed body, a prehensile tail, a tongue that captures and retrieves food, and paired toes are the physical features associated with this species. Individuals at rest are an ash-white to gray with four to five dark bands extending up the sides. A black-encircled spot may rest atop each band. Gravid females maintain a green background with orange markings. Its center of distribution is Madagascar.

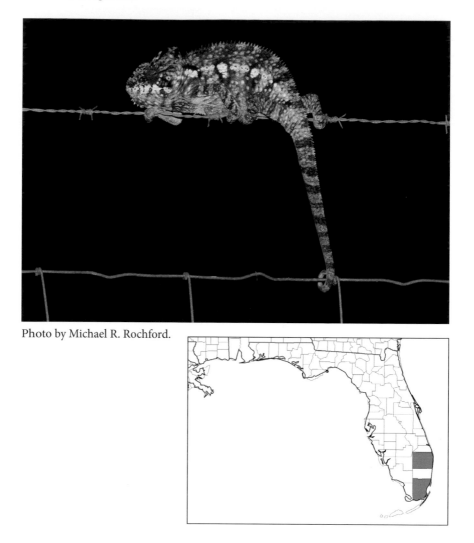

Photo by Michael R. Rochford.

Introduction history and geographic range: The population of this species in the United States was derived from the pet trade. Its existence in Florida City, Miami-Dade County, Florida, since at least 2000, was reported in 2010. In the United States, Oustalet's Chameleon is established in a single avocado grove in southern Florida.

Ecology: Oustalet's Chameleon is a strongly arboreal, diurnal, and cryptic species that has excelled in, yet scarcely dispersed from, the confined grove. Individuals sleep on branches from approximately 72.0 in. above the ground and up. In southern Florida, an average of 42 eggs and maximum of 72 eggs are laid during June–October. The clutch comprises approximately 30% of a female's gravid weight. Eggs take nearly one year to hatch. Consequently, well after apparent extirpation, successive broods await recruitment such that the colony has proven surprisingly resilient despite an onslaught of pet-related and scientific collecting.

Both invertebrates and vertebrates are readily captured with the use of a protractible tongue. Surprisingly large prey can be handled by this species. Raptors, corvids, and racers are the likeliest predators of this species. The Florida City site is associated with very few Brown Anoles, a species it is known to eat. No overlap exists with that population and the nearby Veiled Chameleon population.

Jackson's Chameleon
Trioceros jacksonii (Boulenger, 1896)

Identification: Jackson's Chameleon is a medium-sized, up to 12.0 in., chamaeleonid lizard. Females are smaller than males. Independently moving eyes, laterally compressed body, a prehensile tail, a tongue that captures and retrieves food, and paired toes are the physical features associated with this species. Color and pattern are subject to change by the individual and range from alternating black and gray, yellow-green, and mottled olive. Males are unmistakable with their distinctive three horns; one above the upper jaw, and one above each eye. Females generally do not have these structures. The horns are used in agonistic encounters with other males for access to a female. Its center of distribution is Kenya.

Introduction history and geographic range: The population of this species in the United States was derived from the pet trade dating back to releases on

Photo courtesy of Bill Love.

O'ahu, Hawai'i, in 1972. Animals soon dispersed into the Ko'olau Mountains. By 1995, colonies were detected on Kaua'i and Lāna'i as reported in 1996. It now occurs on Hawai'i and Maui. In California, it was first reported from Redondo Beach, Los Angeles County, and San Diego, San Diego County, in 1996, and in Morro Bay, San Luis Obispo County, in 1997. Accidental release from an enclosure at a commercial breeder/retailer when the business was being raided in a sting operation is the source of the Morro Bay colony. In the United States, Jackson's Chameleon is found in Hawai'i (Hawai'i, Kaua'i, Lāna'i, Maui, and O'ahu) and a portion of coastal California. The species is well established in Hawai'i, especially at midlevel elevations, where it inhabits secondary forests and urban situations, but also in drier habitat of lower elevations.

Ecology: Jackson's Chameleon is a strongly arboreal, primarily diurnal, and cryptic species. Movement is often associated with slow rocking. Receptive females are green. Jackson's Chameleon is ovoviviparous, giving live birth to an average of twelve young approximately six to nine months after mating. A second litter can be produced six months later. Sexual maturity in both sexes is reached in less than one year. The smallest adult males from low elevation sites in Hawai'i are smaller than counterparts from higher elevations in Kenya. If younger, then introduced males are entering the breeding population at a younger age. Individuals will move on wet nights. Habitat suitability will dictate patterns of movement in a way that suggests that in Hawai'i individuals can contribute to some extent to their own dispersal to new target areas. Jackson's Chameleon is an insectivore whose diet includes both native and introduced invertebrates. Prey is captured with the use of a protractible tongue. It has been shown to incorporate native invertebrates in its diet in Hawai'i.

Casque-headed Lizards: Corytophanidae

BROWN BASILISK

Basiliscus vittatus Wiegmann, 1828

Identification: The Brown Basilisk is a typically large, up to 24.0 in., coryto-phanid lizard. Among adults, males are larger than females. Adult males are pale green-brown with a light yellow lateral stripe extending from eye to the vent. A prominent crest on the back of the head and a shorter dorsal crest are also characteristic of adult males. Females and juveniles are brown with a light

Photo by Janson Jones.

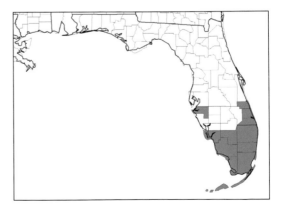

lateral stripe and similar but reduced crests. Its center of distribution is Central America and northern South America.

Introduction history and geographic range: The population of this species in the United States was derived from the pet trade, where it was first detected in Miami, Miami-Dade County, Florida, 1976, and reported in 1983. Subsequent dispersal was rapid along the many interconnecting canals and nearby borrow pits. In the United States, the Brown Basilisk is found throughout much of southern Florida and increasingly along the eastern coast.

Ecology: The Brown Basilisk is associated with edges of freshwater systems bordered with vegetation and trees for cover and climbing. In Florida, ponds and canals with brush and some trees along the shoreline can support this diurnally active species in abundance. Ever wary and known for "walking on water," juveniles especially will escape to the water, where they run bipedally parallel to the shoreline until turning back to land. Air captured under the many folds beneath their toes supports these lizards upon the water. Because of weight, adults are less apt to take to water than are juveniles, preferring to rapidly ascend bushes and trees. A female from Miami was found in the act of nesting on 18 March 2016. She covered the nest chamber after laying six eggs. These eggs were artificially incubated and hatched 66 days later. In captivity, females will lay eggs several times each year. In southern Florida, neonates can be found during summer. In captivity, growth can be rapid, with females maturing within three months. The Brown Basilisk occurs up to Everglades National Park but has not colonized it, perhaps because of both hydroperiod and predators. Disturbed systems, which prevail in southern Florida, are colonized easily, and the species is widespread in the region. Insects and small vertebrates are eaten. Adults readily capture anoles and quickly transform a Brown Anole population to few and large individuals. A confirmed predator of other anoles as well, including the native North American Green Anole, this guild is certainly at risk of predation at low heights or on the ground to lay eggs or occasionally hunt.

Anoles: Dactyloidae

ALLISON'S ANOLE
Anolis allisoni Barbour, 1928

Identification: Allison's Anole is a medium-sized, up to 8.0 in., dactyloid lizard. Among adults, males are larger than females. Background color is green but can quickly change to brown. A displaying male is striking in appearance, having a red dewlap and a bright blue anterior portion of its body. Adult males have a nuchal crest that can be extended and often have a pronounced dorsal crest that extends onto the tail. Its center of distribution is Cuba.

Photo courtesy of Jonathan Losos.

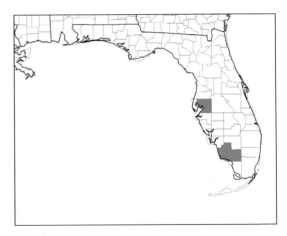

Introduction history and geographic range: Arrival of this species in the United States was human-mediated, where it was reported from Tampa, Hillsborough County, Florida, in 2015 but collected in 2013. However, the first established colony of Allison's Anole in the United States is from Naples, Collier County, where it was captured and reported in 2017. The colony, located on private property, was stated to have been present for many years. The authors of the report first encountered the species at that site in 2014. The status of this species in Hillsborough County is uncertain. In the United States, Allison's Anole is found in Florida.

Ecology: Allison's Anole is a conspicuous mid-trunk-canopy anole. Activity is diurnal. Males are territorial. Aggressive encounters between males can culminate in wrestling with jaw holds. Clutches of one or two eggs are produced. Invertebrates form the largest part of its diet. Little is known about the ecology of Allison's Anole in Florida; however, Allison's Anole overlaps in body size and structural niche with close relatives, the North American Green Anole and the Cuban Green Anole.

NORTH AMERICAN GREEN ANOLE
Anolis carolinensis (Voight, 1832)

Identification: The North American Green Anole is a medium-sized, up to 8.0 in., dactyloid lizard. Among adults, males are larger than females. Individuals will change colors from dark brown to light brown, to olive, to bright green. Adult males have a well-developed pink to red dewlap that is striking in color when distended for communication. Females have a light brown vertebral stripe. Two subspecies are recognized: the Northern Green Anole, *A. c. carolinensis,* and the Southern Green Anole, *A. c. seminolus.* The North American Green Anole is native to the southeastern United States.

Introduction history and geographic range: In the United States, the North American Green Anole is found on several islands of Hawai'i (Hawai'i, Kaua'i, O'ahu, Maui, Moloka'i). The extralimital population of this species in the United States was derived from the pet trade dating back to 1950 on O'ahu,

Photo by Janson Jones.

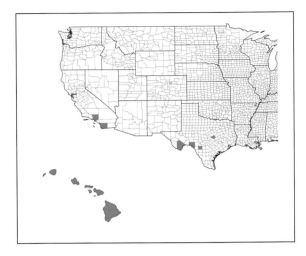

Hawai'i. In California, it is known from San Diego County on the grounds in and around the zoo. It is also reported from Los Angeles County. Accidental introductions may be responsible for records in ten Texas counties; however, only individuals seen in Brewster County appear to be derived from intentional introductions.

Ecology: The North American Green Anole can be found in a wide range of habitats with shrubs and trees. It is diurnal. Although strongly arboreal, it will also forage close to the ground. In Hawai'i, this species is most frequently associated with human habitations, residences, gardens, oceanside resorts. Females and juveniles are less easily seen than territorial males, which often prominently perch on trunks of trees.

Males are highly territorial and will often lock jaws in wrestling matches. In southern Florida, multiple clutches of up to ten single-egg clutches are laid from April through September. Sexual maturity is reached in about seven or eight months. The North American Green Anole is primarily insectivorous and was Hawai'i's first and only diurnal insectivorous arboreal lizard for 30 years. Subsequent arrivals of the Brown Anole and the Knight Anole represent predatory and competitive members of the lizard assemblage in Hawai'i and Florida. In syntopy, the Brown Anole is often far more abundant than the North American Green Anole, and it is far more apt to eat North American Green Anoles than the other way around. Observationally, where the Knight Anole is present in southern Florida, the Brown Anole remains closer to the ground, and the North American Green Anole appears to persist more successfully. In light of similar aspects in ecology, the North American Green Anole is subject to potential predation and competitive interactions with the Puerto Rican Crested Anole. It is also subject to hybridization with the Cuban Green Anole.

Hispaniolan Green Anole
Anolis chlorocyanus Duméril and Bibron, 1837

Identification: The Hispaniolan Green Anole is a medium-sized, up to 8.0 in., dactyloid lizard. Among adults, males are larger than females. The dorsum is bright green but varies with the state of the individual. Dorsal stripes may be present on females and juveniles. Among adults, the blue-black dewlap is pronounced in the male and small in the female. Its center of distribution is Hispaniola.

Introduction history and geographic range: Arrival of this species in the United States was through the pet trade, where it was reported from Miami, Miami-Dade County, Florida, in 1988. The colony no longer existed after construction at the site for a train station. Subsequent introductions were also associated with the pet trade. A second colony was reported from Parkland, Broward County, in 1994, where it had been present since 1987. In the United States, the Hispaniolan Green Anole is found in southern Florida. The Miami-Dade County population and one reported for Martin County apparently did not survive. The latter population was eliminated by a freeze; however, a small but persistent population in Broward County has been in existence since 1987.

Photo courtesy of Jonathan Losos.

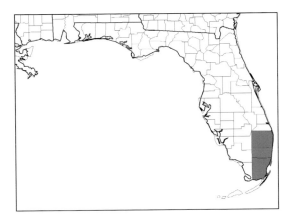

Ecology: In Broward County, the Hispaniolan Green Anole prefers large trees where it is typically found from mid-trunk to the tree canopy and large branches. On buildings and trees, individuals were typically found at or above 6 ft. The population size was small and much smaller than that of the syntopic Hispaniolan Stout Anole. Activity is diurnal. Males are territorial, and females produce multiple clutches; however, the length of the breeding season in southern Florida remains to be determined. Its diet is thought to be composed primarily of arthropods. From what little is known of its ecology in southern Florida, the Hispaniolan Green Anole seems most similar to the native North American Green Anole in habits and could potentially be a competitor. However, removal of large trees from the Broward County site has negatively impacted an already small population placing the existence of this otherwise static colony in doubt.

PUERTO RICAN CRESTED ANOLE

Anolis cristatellus Duméril and Bibron, 1837

Identification: The Puerto Rican Crested Anole is a medium-sized, up to 7.5 in., dactyloid lizard. Among adults, males are larger than females. Background color is dark brown but can change to greenish gray. A pair of rust-colored paravertebral stripes may be present in males. The color of the dewlap of the adult male is reddish orange and greenish in the middle. Adult males have a nuchal crest that can be extended and it often extends onto the tail. Females and juveniles have a light mid-dorsal stripe. Its center of distribution is Puerto Rico and the U.S. Virgin Islands.

Introduction history and geographic range: Arrival of this species in the United States was human-mediated, where it was reported from Key Biscayne, Miami-Dade County, Florida, in 1975. Subsequent introductions on southern mainland Florida were deliberate. In the United States, the Puerto Rican Crested Anole is

Photo courtesy of Janson Jones.

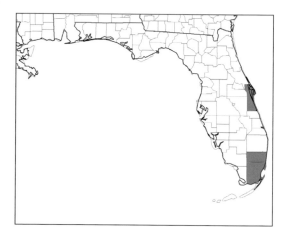

found in Florida (Brevard and Miami-Dade counties). The status of the Brevard County population is uncertain.

Ecology: In southern Florida, the Puerto Rican Crested Anole is typically found in areas with partial to extensive shade, where individuals occupy the trunk-ground niche. Adult males, especially, can be seen on higher perches displaying and fighting other males during the breeding season. Juveniles and females are most often seen near the ground and among cover. Populations of this species are not especially expansive, but individuals can be very abundant where they occur. Activity is diurnal. Males are territorial. Aggressive encounters between males can culminate in wrestling with jaw holds. Multiple clutches of one, occasionally two, eggs were produced between March and November. Sexual maturity is presumed to be reached within one year. A wide range of arthropods is eaten, especially ants and beetles. One individual defecated a Brahminy Blind Snake. Racers are likely predators of this lizard. Despite several decades in southern Florida, this species has remained relatively localized in its historic introduction sites. The Puerto Rican Crested Anole and the Brown Anole are ecologically similar to one another in several respects. In contact with one another, the Puerto Rican Crested Anole socially dominates the Brown Anole; however, the extent of shade enforces separation of the two species. In light of its diet and perch preferences, the Puerto Rican Crested Anole is a potential predator and competitor of the North American Green Anole, the Cuban Green Anole, and possibly the Bark Anole.

HISPANIOLAN STOUT ANOLE
Anolis cybotes Cope, 1862

Identification: The Hispaniolan Stout Anole, aka the Large-headed Anole, is a medium-sized, up to 8.5 in., dactyloid lizard. Among adults, males are larger than females. Background color is generally light to dark brown. Among adult males, a light lateral stripe is usually present, and the color of the dewlap is a shade of yellow. The dewlap and the head are large in adult males and result in an overall bulldog-like appearance. An extendable nuchal and dorsal crest are present. Females have a light shoulder spot and a light mid-dorsal and lateral stripe. Its center of distribution is the Caribbean.

Photo by Suzanne Collins.

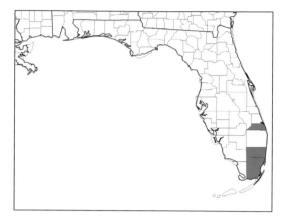

Introduction history and geographic range: Arrival of this species in the United States was intentional, where it was reported from Miami, Miami-Dade County, Florida, in 1973. Subsequent introductions in Florida were also deliberate. In the United States, the Hispaniolan Stout Anole is found in southeastern Florida.

Ecology: In southern Florida, the Hispaniolan Stout Anole occupies a variety of perch types in disturbed areas but closer to the ground like the Puerto Rican Crested Anole and the Brown Anole. Adult males often perch facing head-down on tree trunks with their heads extended nearly parallel with the ground. Activity is diurnal. Males are strongly territorial. Multiple clutches of one, occasionally two, eggs are produced; however, the length of the egg-laying season is not known. Its diet is omnivorous and includes small vertebrates. The Hispaniolan Stout Anole is much more abundant than the Hispaniolan Green Anole, where they co-occur in a Broward County site. Although the Hispaniolan Stout Anole does not appear to disperse readily on its own from introduced sites, intentional introductions could place it in direct contact with the North American Green Anole, which it could negatively impact socially or through predation. The same could be true of the other smaller anoles.

BARK ANOLE
Anolis distichus Cope, 1862

Identification: The Bark Anole is a small, up to 5.0 in., dactyloid lizard. Among adults, males are larger than females. Background and dewlap color vary, depending on the extent of intergradation that exists between two forms. The Florida Bark Anole, *A. distichus floridanus,* is grayish in background color, and adult males have a light yellow dewlap. This form may be native to Florida. The Green Bark Anole, *A. d. dominicensis,* is yellowish green in background color, and adult males have a yellow dewlap with a distinct orange spot in the center.

Photo by Suzanne Collins.

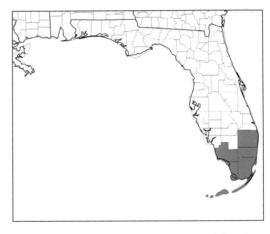

Most of what one sees is an intergrade of the two forms. Its center of distribution is the West Indies. Detection of this species in the United States was first reported from Miami, Miami-Dade County, Florida, in 1948, and described as the Florida Bark Anole.

Introduction history and geographic range: Arrival of the Green Bark Anole in the United States, presumed to have been incidental through trade with Hispaniola, was first reported from Miami, Miami-Dade County, Florida, in 1966. Other forms of this species, unsuccessful in their colonization, have been reported elsewhere in Florida. In the United States, the Bark Anole is found in southern Florida.

Ecology: The Bark Anole occupies the mid-trunk structural niche, and in disturbed habitat it is found most often on large smooth-barked trees, such as *Ficus* trees and smooth-trunked palms, such as royal and foxtail palms. A frightened Bark Anole will rapidly climb up the trunk and very soon after return to its preferred trunk height. Activity is diurnal, and individuals prefer a microhabitat of semi-shaded, broken filtered light. Activity is bimodal during hot weather and unimodal in the winter. Multiple clutches of single eggs are produced from March through October. Hatchlings are seen from July through November, and most born in the summer can reproduce the following season. Ants are its primary prey, captured by sitting and waiting on the lower section of the tree trunk. More foraging occurs in the morning than in the afternoon. The Cuban Green Anole is a confirmed predator of this species. Despite its preference for shadier microhabitat than the North American Green Anole, when in syntopy the latter species will move to sunnier positions.

Knight Anole
Anolis equestris Merrem, 1820

Identification: The Knight Anole, or Chipojo, is a large, up to 20.0 in., dactyloid lizard. Among adults, males are larger than females. A sharply angular casqued arrowhead-shaped head is a distinctive trait in this species. Adults are bright grassy green with a yellow shoulder stripe; however, background color can change to darker green or a dark mustard brown. Juveniles bright green with transverse cream-colored stripes. Both sexes have a light pink dewlap. Its center of distribution is Cuba.

Photo courtesy of Janson Jones.

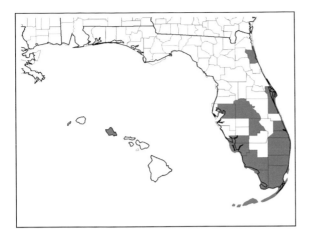

Introduction history and geographic range: Arrival of this species in the United States was human-mediated, where it was reported from southern Florida in 1957. A population in Coral Gables, Miami-Dade County, Florida, was derived from intentional release and reported in 1966. In Hawai'i, it was first detected on O'ahu in 1981 and reported in 1996. The source of its introduction was release of pets. In the United States, the Knight Anole is well established throughout much of southern Florida and portions of central Florida. It is incidentally, through lawn care and ornamental plant shipments, and intentionally moved about. In Hawai'i, the Knight Anole is established in two sites on O'ahu.

Ecology: In Florida, the Knight Anole is most closely associated with large branches and the trunks of large trees. In Florida, this species is most abundant in sites analogous to secondary growth whereby large and medium-sized trees with canopy are interspersed in a way that provides plenty of opportunity to bask in direct sun as needed. Consequently, tree-lined residential developments and county parks provide habitat much to their liking. Most adults and subadults are seen at heights of at least 6 ft. above the ground but can still be seen hunting close to the ground. Perches head-down with head jutting out on the trunks of trees, such as black olive, mahogany, live oak, and wild tamarind, and *Bischofia* is a typical site from April through October when individuals are most active. Hatchlings and young-of-the-year are nearly always found within 2 ft. of the ground in bushes. Activity is diurnal, and individuals are most apt to be actively hunting and basking in ambient temperatures greater than 84°F and in months whose average high was at least 86°F. In temperatures exceeding 90°F, they are found in deep shade. During the cooler drier months, individuals are seldom seen, spending much of their time nearer the tops of trees. Where they have not been heavily collected, individuals are easily approached. A frightened Knight Anole will slowly or rapidly climb up the trunk and hide from its pursuer.

Courting males will fight from March through August, but especially in May. Lacerations on the body, especially around the heavily casqued head, are common. Tails are lost in some of these encounters. Mating takes place conspicuously on large branches or tree trunks from March through September. Multiple clutches are produced at least in June and July and likely over a longer time. Males are mature in 13 months, and females in 7.5 months. Population turnover in a Homestead, Miami-Dade County, Florida, population occurred in about five years. The Knight Anole is an omnivore that both sits and waits for its food and actively forages. Various fruits, berries, flowers, leaves, and sap are eaten. Blue-grey Gnatcatchers and Purple Martins are eaten. Cuban Treefrogs are dug out of epiphytes and eaten. Brown Anoles and Indo-Pacific House Geckos are also eaten. Individuals will explore the confines of palm boots and tree cavities for food as well as await moving prey to pass near them. Prey are caught in trees and on the ground by individuals pouncing from a tree.

Depredations of the Knight Anole place a wide range of native and exotic vertebrates at risk. The Knight Anole is a likely predator of the Cuban Green Anole. The North American Green Anole is often more abundant in the presence of the Knight Anole than when it is alone with the Brown Anole. A potential "*T. rex* of the trees," the Knight Anole should be looked at as a, if not the, driving force in anoline assemblage structure in Florida. The Northern Mockingbird, Blue Jay, and racer may be predators of young Knight Anoles and the former two species may also be potential competitors with adults for food.

JAMAICAN GIANT ANOLE
Anolis garmani Stejneger, 1899

Identification: The Jamaican Giant Anole is a large, up to 12.0 in., dactyloid lizard. Among adults, males are larger than females. Background color is bright green and can change to dark brown. A short saw-toothed crest is present continuously from the nape to the tail. Adult males have a brownish-yellow dewlap. Its center of distribution is the West Indies.

Introduction history and geographic range: Arrival of this species in the United States was human-mediated where it was reported from South Miami, Miami-Dade County, Florida, in 1983. The population had been in existence since at least 1975. In the United States, the Jamaican Giant Anole occurs in a long-established and highly localized neighborhood in Miami-Dade County, Florida. Colonies in Lee and Martin counties did not survive.

Ecology: The Jamaican Giant Anole is a large branch anole, often perching head-down within a few feet of the ground. The South Miami site consists of disturbed habitat with large shade trees. Activity is diurnal, and, during the summer months, adults perch conspicuously on tree trunks and large branches

Photo courtesy of Janson Jones.

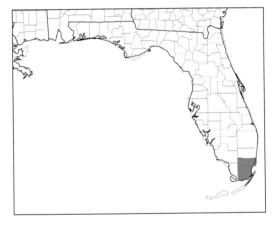

from a few to 40 feet above the ground. The species is active throughout the year, especially in the early afternoon. Activity in Florida is most apparent from fall through spring. This species has been reported to lay four-egg clutches through the summer. A female collected in September contained one shelled egg and an enlarged ovarian follicle, suggestive of multiple single-egg clutches. The Jamaican Giant Anole is an omnivore: the Brown Anole, *Ficus* fruits, and Hibiscus leaves are reported in its southern Florida diet. The static and confined nature of this population is enigmatic. Ecologically, it most closely resembles the Knight Anole, a species abundant in areas surrounding the Jamaican Giant Anole colony. It remains to be seen why the Jamaican Giant Anole persists in this neighborhood and cannot expand beyond it.

CUBAN GREEN ANOLE

Anolis porcatus Gray, 1840

Identification: The Cuban Green Anole is a medium-sized, up to 8.5 in., dactyloid lizard. Among adults, males are larger than females. The skull is long and pointed and rugose. Two prominent frontal ridges run lengthwise. The background color is bright green with varying degrees of creamy vermiculation. The dewlap is pinkish to purple-mauve in color. The Cuban Green Anole is closely related to the North American Green Anole, and they superficially resemble one another. Its center of distribution is Cuba. Hybrids of the two species occur in southern Florida.

Photo courtesy of Jonathan Losos.

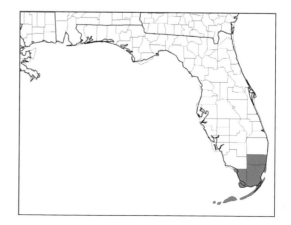

Introduction history and geographic range: Arrival of this species in the United States was human-mediated, where it was reported from Key West, Monroe County, Florida, in 1945. In 1997, a confirmed colony was reported in South Miami, Miami-Dade County, that had been in existence since 1987 and one in North Miami, Miami-Dade County, Florida, that had been in existence since 1991. The derivations of those two populations and subsequent ones in Coconut Grove and Kendall, both of Miami-Dade County, are unknown. In the United States, the Cuban Green Anole occurs in multiple sites in Florida (Miami-Dade County).

Ecology: In Florida, the Cuban Green Anole occupies the mid-trunk–canopy niche, where it is seldom seen within 6 ft. of the ground. It is diurnal in activity and active throughout the year. Multiple single-egg clutches are produced, as determined by examination of females captured in July. The Cuban Green Anole is an omnivore. Fruits and flowers are eaten. Much of its invertebrate prey is small in body size and comprises mostly flies, ants, beetles, and spiders; however, it also consumes the Bark Anole and Brown Anole. Having evolved with two other Cuban anoles that are also in Florida, the Cuban Green Anole would be expected to fare much more successfully than the North American Green Anole when in syntopy with the Knight Anole and the Brown Anole. The Cuban Green Anole presents a risk to the North American Green Anole, not only through predation and potential competition for perches and food, but also through hybridization. If so, the likeliest scenario would be social dominance of male Cuban Green Anoles and mating with female North American Green Anoles by the larger-bodied Cuban Green Anole. The potential for competition and hybridization with, and predation on, the native North American Green Anole in disturbed habitats elevates the threat by this species to a level of concern that should not be underestimated. For this reason, collections of series of these two species should be made from the Florida Keys at least as far north as Ft. Myers, Lee County, to ascertain the extent to which the Cuban Green Anole has dispersed and the direction of potential hybridization.

Brown Anole
Anolis sagrei Duméril and Bibron, 1837

Identification: The Brown Anole is a medium-sized, up to 8.0 in., dactyloid lizard. Among adults, males are larger than females. The background color of an adult male varies from uniform tan to shades of gray with lateral and paravertebral black stripes. Irregular vertical stripes or gold stippling may be present along the sides of its body. A displaying male can extend its nuchal and vertebral crest, adding nearly 20% to its height. The vertebral crest usually extends only slightly past the rear legs, although males with a caudal crest are seen

Photo courtesy of Carlos Nieves.

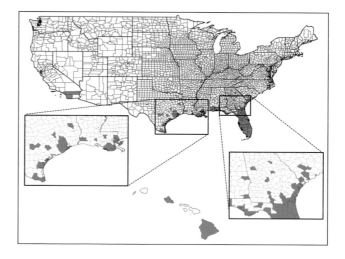

occasionally. The dewlap is red, interspersed with rows of small black dots, and is bordered in bright yellow. Females and juveniles are grayish or some shade of brown in background color. A scalloped pattern borders a light vertebral stripe. Its center of distribution is Cuba.

Introduction history and geographic range: Arrival of this species in the United States was reported from the Florida Keys, Monroe County, Florida, in 1887. This species was not reported from the Florida mainland until nearly 100 years after the initial Florida record. The status of being native to the lower Florida Keys cannot be ruled out; however, subsequent dispersal to the mainland is best explained by human mediation. In Alabama, the Brown Anole was first reported from Opelika, Lee County, in 2007. Derived through the ornamental plant trade, the population remains extant at the time of this writing. In California, it was first reported from Vista, San Diego County, in 2014. The derivation of this colony was associated with a 2012 shipment of plants from Hawai'i. In Georgia, it was first reported from Valdosta, Lowndes County, in 1995. Annual surveys beginning in 1983 did not detect the Brown Anole until 1987. The colony was thriving by 1988 and still in existence in 1994. In Hawai'i, it was first reported on O'ahu in 1996, where it had been since 1980. It is unclear whether this population was derived incidentally from nursery shipments or from the pet trade. In Louisiana, it was first reported from New Orleans, Orleans Parish, in 1990, where it was first collected in 1988. In South Carolina, it was first reported from rest areas on I-95, Colleton County, I-95, Jasper County, and I-26, Orangeburg County, in 2006, where it had been collected in 2005. Colonies persisted through 2009 but may no longer exist since a 2010–2011 frost. The method used by the authors to study these South Carolina colonies was commendable. In Texas, it was first reported from Houston, Harris County, in 1987, where it had been collected in 1985. Also, in 1987, from collections made in 1986, populations were found at a nursery in each of three sites: Brownsville and Harlingen (Cameron County) and San Antonio (Bexar County). All three populations had been in existence for at least three years prior to the time collections were made. In the United States, the Brown Anole occurs throughout most of Florida and in isolated colonies in Alabama, California, Georgia, Hawai'i, Louisiana, South Carolina, and Texas. On the mainland, it remains a species of southern latitudes.

Ecology: The Brown Anole occupies the ground–mid-trunk niche, where it is most often seen from the ground to about 3 ft. above the ground, and this is especially true of females and juveniles. In areas without the Knight Anole, males are more apt to ascend higher above the ground. Conversely, where the Knight Anole is commonly seen, even large males of the Brown Anole stay closer to the ground. Apart from rocks and vegetation, the Brown Anole will climb buildings and thrives in sunny, relatively open habitat with structure such as bushes to hide in or climb. To that end, urbanization and canopy-clearing effects of hurricanes are to the benefit of this species.

It is primarily diurnal in activity and will often bask on sidewalks, the ground along hedges, and vertical structures. However, individuals will often actively forage in light rain, and some individuals will forage into the night alongside geckos near lights on buildings. Activity is essentially continuous throughout the year, with cold spells depressing activity. Copulation has been observed from June through September in Florida. On the Florida Keys, eggs are produced singly every two weeks throughout the year unless conditions are too cool or dry. In southern Florida and Hawai'i, the reproductive cycle of both sexes is strongly associated with day length. Multiple single-egg clutches are possible from January through October, peaking April through July in southern Florida, and January through September in Hawai'i. Neonates are evident from June through October in Miami, Florida, and from June through September on O'ahu.

The Brown Anole is primarily an insectivore but will eat other lizards, such as the North American Green Anole, conspecifics, and possibly the Bark Anole. Ecologically versatile, the Brown Anole can quickly colonize many sorts of habitats. This species is socially dominated by the Crested Anole. The Brown Anole is also more apt to eat juveniles of the North American Green Anole than the other way around, thereby negatively impacting a species it can quickly outnumber. The Brown Anole can be an abundant prey item for many species, including the Ring-necked Snake, Red Cornsnake, *Pantherophis guttatus*, Everglades Racer, *Coluber constrictor paludicola*, Southern Black Racer, *C. c. priapus*, North American Green Anole, Giant Ameiva, Nile Monitor, Cuban Treefrog, Cuban Green Anole, Jamaican Giant Anole, Knight Anole, Brown Basilisk, and Northern Curly-tailed Lizard.

True Geckos: Gekkonidae

BIBRON'S THICK-TOED GECKO
Chondrodactylus bibronii (Smith, 1845)

Identification: Bibron's Thick-toed Gecko is a small, up to 5.0 in., gekkonid lizard. Among adults, males are larger than females. Individuals are stocky in build, mottled in brown and black, and are strongly tuberculate. The venter is white. The overall pattern is lighter at night when individuals are active. Its center of distribution is Africa.

Introduction history and geographic range: The population of this species in the United States was derived from the pet trade. It was reported in Bradenton, Manatee County, Florida, in 1999, where it had been present since the 1970s.

"Bibron's Thick-toed Gecko (Pachydactylus bibronii)" by Bernard Dupont is licensed under CC BY-SA 2.0. https://www.flickr.com/photos/65695019@N07/50050421383.

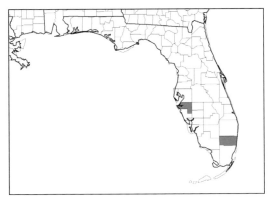

Ecology: In Bradenton, Florida, Bibron's Thick-toed Gecko is found on buildings and less so on nearby trees. Nocturnal in habits, individuals are most frequently encountered in a head-down position on walls as they hunt. Captives will lay several two-egg clutches during the warm months of the year. Invertebrates and small vertebrates make up its diet. Very little is known about this geographically restricted species in Florida. Its likeliest potential competitors are the Cuban

Treefrog and the Tokay Gecko—a competitive and predatory relationship that could be bi-directional. Although larger than the hemidactyline geckos upon which it could depredate or potentially outcompete for food and ideal spots near lights, Bibron's Thick-toed Gecko remains for reasons unknown a comparatively poor colonizer in southwestern Florida. Stochastic reasons or possibly negative impacts from other exotic species could place this species at a high risk of extirpation.

ROUGH-TAILED GECKO
Cyrtopodion scabrum (Heyden, 1827)

Identification: The Rough-tailed Gecko is a small, up to 4.5 in., gekkonid lizard. Individuals are slender in build, mottled in brown on a tan background. Tubercles are present on the dorsum, sides, and tail. The tail is narrow and long compared with those of the exotic *Hemidactylus* species. Its center of distribution is the Middle East.

Introduction history and geographic range: Arrival of this species in the United States was first detected in 1983, derived from shipping at the port of Galveston, Galveston County, Texas, and reported in 1984. Because of its presence in ports and other types of human settlements, incidental human-mediated dispersal by way of cargo was a likely agent of its dispersal to the Port of Galveston. The

"*Cyrtopodion scabrum*; Keeled Rock Gecko" by Omid Mozaffari. https://calphotos. berkeley.edu/cgi/img_query?seq_num=192721&one=T.

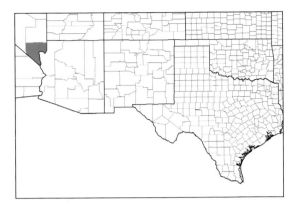

age of the colony was not known. However, when first discovered at Galveston, the Rough-tailed Gecko was outnumbered in captures of the Mediterranean Gecko by 1:3. The subsequent dominance of the Rough-tailed Gecko in Galveston could mean that the initial report in Galveston was of a recently established colony. In Nevada, it was first reported from Apex Dunes, Clark County, in 2017, where it had been collected in 2015 and seen in 2016. In the United States, the Rough-tailed Gecko occurs in two very different habitats and climates, which demonstrates potential for geographically broad colonization.

Ecology: Individuals are most frequently encountered on buildings and other human-made structures; however, in Nevada, the Rough-tailed Gecko was found in disturbed habitat. Activity is nocturnal. Hatchlings have been seen in September. This species is an insectivore. Allopatric populations of the Rough-tailed Gecko and the Mediterranean Gecko eat similar prey; however, their prey differ markedly where the two geckos co-occur. Neither the Mediterranean Gecko nor the Common House Gecko persists alongside the Rough-tailed Gecko. A 1996 report states that a single Common House Gecko was collected in 1988 in the vicinity of the Rough-tailed Gecko population. No other individuals were found in seven subsequent trips in 1989 and 1990.

MUTILATING GECKO
Gehyra mutilata (Wiegmann, 1834)

Identification: The Mutilating Gecko is a small, up to 2.5 in., gekkonid lizard. Adults are stout in form. The tail is constricted at its base and dorsoventrally flattened. Depending on the circumstances, the dorsum may be black with light speckling, dark brown, or nearly white. A thin light ocular stripe and light banding on the tail may be present. Its venter is yellowish. Four large chin shields are present. Its center of distribution is Southeast Asia.

"Photo 78388818" by desertnaturalist is licensed under CC BY 4.0. https://inaturalist.ca/photos/78388818.

Introduction history and geographic range: The Mutilating Gecko is presumed to have colonized the Hawaiian Islands as a stowaway with the indigenous people of Hawai'i. In the United States, the Mutilating Gecko is found on the islands of Hawai'i, Kaho'olawe, Kaua'i, Lāna'i, Maui, Moloka'i, Ni'ihau, and O'ahu.

Ecology: Found in forests and buildings, individuals are nocturnal in their activity. Two nearly round and adhesive hard-shelled eggs are laid at a time. Incubation lasts approximately two months. This species is an insectivore. In urban areas of Hawai'i, the Mutilating Gecko is subject to displacement by the Common House Gecko.

Golden Gecko

Gekko badenii Szczerbak and Nekrasova, 1994

Identification: The Golden Gecko is a medium-sized, up to 7.0 in., gekkonid lizard. Among adults, males are larger than females. Hemipenal bulges are apparent on adult males. Adults are slender in form. As per their common name, individuals are light gold in color with narrow light bands across the back; however, the color in females is more subdued than in males. The venter is uniform white or cream in color. Its center of distribution is Vietnam.

"Gekko badenii" by Heroinabspeutzer is licensed under CC BY-SA 4.0.
https://commons.wikimedia.org/w/index.php?curid=38087010.

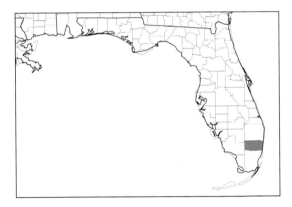

Introduction history and geographic range: Populations of this species in the
United States were derived from the pet trade and first detected in 2008 in Hol-
lywood, Broward County, Florida, and reported in 2011. In the United States,
the Golden Gecko is found in highly localized circumstances associated with an
animal importer facility in Broward County.

Ecology: The Golden Gecko is found in heavily disturbed habitat. It is nocturnal in activity. Males are territorial. Two adhesive hard-shelled eggs are laid at
a time. Incubation lasts between two and nearly three months depending on
temperature. This species feeds on both nectar and insects. Effects on or by the
Golden Gecko with other vertebrates remain unknown.

TOKAY GECKO

Gekko gecko (Linnaeus, 1758)

Identification: The Tokay Gecko is a large, up to 12.0 in., gekkonid lizard. Among adults, males are larger than females. This species is distinctive with its tuberculate sides and orange and red spots on a mauve to blue background. Its venter is pale and flecked in orange and red. The tails of juveniles are banded in black. The Tokay Gecko is an Old-World tropical species.

Introduction history and geographic range: Populations of this species in the United States were derived from the pet trade. In Florida, it was first reported in Broward and Miami-Dade counties in 1983, where it had been present in disparate localities since the 1960s. In Hawai'i, a population was first detected in Lanikai, O'ahu, around 1980 and reported in 1996. Prior to the establishment, individuals were brought to Hawai'i as pets in the 1960s and early 1970s by servicemen returning from Southeast Asia. Subsequent dispersal through intentional release is common. In the United States, the Tokay Gecko is found primarily in the southern half of peninsular Florida and on O'ahu in Hawai'i.

Photo by Suzanne Collins.

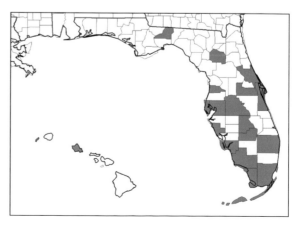

Ecology: The Tokay Gecko prefers large trees with cavities and can be successful on buildings with refuges. In southern Florida, *Ficus* trees are a favored habitat. In northern Florida, activity is curtailed in the winter. The Tokay Gecko is primarily nocturnal but will occasionally be seen in the open near its retreat on sultry days. In southern Florida, males call during winter–spring. In northern Florida, calling has been heard during May–June. The call by territorial males is a loud and sudden "Eh' Awww" or "Gek' Oh." Females lay more than one clutch of eggs, two at a time, at least during May–September. In Hawai'i, egg-laying can occur year-round. Eggs are carefully fastened by the female to a vertical substrate, often communally and repeatedly. A large male is often seen in attendance of these communal egg plaques. Snakes, such as the Red Cornsnake, geckos, and the Cuban Treefrog are included in its diet, with roaches, beetles, and moths especially common, the last associated mostly with geckos foraging near lights. The Tokay Gecko presents a predatory threat to small vertebrates and invertebrates in disturbed tropical hardwood hammocks and oak-palm hammocks in southern Florida. Among the exotic herpetofauna, the Cuban Treefrog is the closest ecological analogue to the Tokay Gecko and a potential predator of its young.

COMMON HOUSE GECKO

Hemidactylus frenatus Duméril and Bibron, 1836

Identification: The Common House Gecko is a small, up to 4.5 in., gekkonid lizard. This species is gray in dorsal color with faint but dark longitudinal stripes. Its venter is white. Its tail is round in cross section and easily broken. Its center of distribution is Southeast Asia.

Photo by Suzanne Collins.

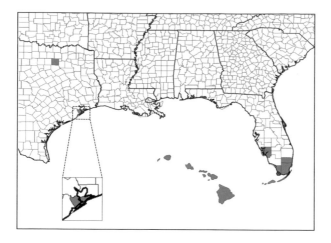

Introduction history and geographic range: The earliest populations of the
Common House Gecko were in Hawai'i, with their dispersal associated with
movements of cargo among the Pacific Islands during or soon after World War
II. In Texas, it was first reported from the Dallas Zoo in Dallas, Dallas County,
in 1990, where it had been introduced for pest control in the 1970s. In Florida,
it was first reported on Key West and Stock Island, Monroe County, in 1994,
where its introduction was associated with the pet trade. Subsequent dispersal
of the Common House Gecko can be very successful by individuals moved
on vehicles, plants, and through the pet trade, where it is purchased as a pet
and as a source of food for lizard-eating pets. In the United States, the Common House Gecko is found in southern Florida, throughout Hawai'i (Hawai'i,
Kaho'olawee, Kaua'i, Lāna'i, Maui, Moloka'i, and O'ahu), and in two locations
in Texas.

Ecology: The Common House Gecko thrives on buildings and in disturbed forest. It is primarily nocturnal. Individuals are highly vocal. Females lay up to four
clutches each year in southern Florida, and individuals reach sexual maturity
within one year of life. The Common House Gecko is chiefly an insectivore. In
Florida, the Ringed Wall Gecko is a confirmed predator of the Common House
Gecko and the Asian Flat-tailed House Gecko. The Tokay Gecko and Cuban
Treefrog are likely predators of this species as well. It is a superior competitor
of the Indo-Pacific House Gecko and the Asian Flat-tailed House Gecko and is
in turn replaced by the Tropical House Gecko. In Hawai'i, the Common House
Gecko is the most successful gecko, where it negatively impacts the Indo-Pacific
House Gecko, Stump-toed Gecko, Mutilating Gecko, and Mourning Gecko. The
Common House Gecko fares poorly in syntopy with the Rough-tailed Gecko.
In 1996, from a 1988 collection, the Common House Gecko was reported at the
same location in Galveston, Texas, as the Rough-tailed Gecko. Two years of
monitoring the site revealed no other individuals of the Common House Gecko.

INDO-PACIFIC HOUSE GECKO

Hemidactylus garnotii Duméril and Bibron, 1836

Identification: The Indo-Pacific House Gecko is a small, approximately 4.5–5.0 in., gekkonid lizard. Individuals of this all-female species are easily identified by what appears as a flesh-colored dorsoventrally compressed body. Its venter is yellow. The tail is flat, saw-tooth-edged, and orange underneath. Endolymphatic sacs of adults are often chalky-white in appearance and filled with calcium deposits. Its center of distribution is Southeast Asia.

Introduction history and geographic range: Arrival of this species in the United States was probably through dispersal by Polynesians to Hawai'i. The Indo-Pacific House Gecko was first known from the mainland in Miami and Coconut

Photo by Suzanne Collins.

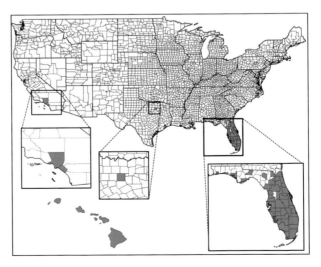

Grove, Miami-Dade County, Florida, in 1966, where it had been present since the early 1960s. In California, it was first reported from Los Angeles and Orange counties in 2015, where individuals had been captured in 2013 and seen since 2009. In Georgia, it was first reported from Brunswick, Glynn County, in 2011, where it was collected in 2006. In Texas, it was first reported from Dallas, Dallas County, in 1996. In the United States, the Indo-Pacific House Gecko is found in Florida, where it inhabits vegetation and buildings of disturbed habitat. Colonies are less expansive in Hawai'i, where it is found in urban areas and forest, in Georgia, and in Texas, it is found at the Dallas Zoo.

Ecology: The Indo-Pacific Gecko is most often encountered on buildings with lights and also on nearby trees. Individuals are primarily nocturnal. More so than maintain territories like other congeners, Indo-Pacific House Geckos wander about buildings, often fighting with one another and with more sedentary congeneric species. Females lay up to three clutches per year of one or two eggs. Adults reach sexual maturity within one year of life. Flies and hymenopterans were found to be its favorite foods in Everglades National Park. The Red Cornsnake, Knight Anole, and Cuban Treefrog are among its predators. It shares a similar diet with the Tropical House Gecko and rapidly disappears entirely or nearly so from buildings soon after colonization by its superior competitor and potential predator. The presence of the Cuban Treefrog slows down the process and often results in incomplete replacement. In urban areas of Hawai'i, the Indo-Pacific House Gecko is negatively impacted by the Common House Gecko. Adult body sizes of the Indo-Pacific Gecko are larger in the presence of the Cuban Treefrog, presumably the result of more food for fewer geckos and some level of protection from its predator associated with larger body size.

Tropical House Gecko, aka Wood Slave
Hemidactylus mabouia (Moreau de Jonnès, 1818)

Identification: The Tropical House Gecko, aka Wood Slave, is a small, approximately 4.5–5.0 in., gekkonid lizard. This species is easily identified by dark chevrons along a light to dark dorsum. Individuals can fade to almost entirely white. The venter is white. Tubercles are scattered on the body, and dark bands are present on the tail. Its tail is round in cross section, also tuberculate, and easily broken. Its center of distribution is Africa.

Introduction history and geographic range: Arrival of this species in the United States was through incidental human-mediated dispersal from South America to Crawl Key, Monroe County, Florida, probably since the early 1980s, and reported in 1991. Soon thereafter, it was detected on mainland Florida. In Texas, it was first reported from Brownsville, Cameron County, in 2014. Individuals were captured on buildings within a circle about 218 yd. in diameter. The

Photo by Suzanne Collins.

source of this colony was thought to be transportation of bricks from Mexico or ornamental plants from Florida. Subsequent dispersal occurs intentionally by humans and incidentally on vehicles and especially on ornamental plants, especially palms. In the United States, the Tropical House Gecko is found throughout much of Florida and in Brownsville, Texas.

Ecology: The Tropical House Gecko can be abundant on buildings with light-attracted invertebrate prey and on trees in disturbed habitat, especially *Ficus*, and on palms with much thatching. On buildings, individuals can be found from the eaves to the ground. Their dorsal pattern is well suited for camouflage on tree trunks and branches, although juveniles and adults will both descend to the ground to forage near the base of the tree.

This species is primarily, but not exclusively, nocturnal in activity. Females lay several clutches of eggs, one or two at a time, throughout the year in southern

Florida. Adults reach sexual maturity within one year of life. On lighted build-ings, the Tropical House Gecko feeds on a wide range of invertebrate prey, espe-cially spiders, flies, and moths. It will also capture and eat small geckos. The Red Cornsnake, Tokay Gecko, and Cuban Treefrog are among its predators. This often-abundant gecko is a superior competitor and potential predator of other hemidactyline and sphaerodactyline geckos on buildings. The Ashy Gecko is replaced by the Tropical House Gecko on the Florida Keys. Upon colonization, replacement of the Indo-Pacific Gecko occurs rapidly and results in many more Tropical House Geckos than previous inhabitants. Replacement of the Indo-Pacific Gecko is delayed and even incomplete in the presence of the Cuban Treefrog. Nonetheless, the Tropical House Gecko is the dominant exotic gecko in Florida.

SRI LANKAN SPOTTED HOUSE GECKO
Hemidactylus parvimaculatus (Deraniyagala, 1953)

Identification: The Sri Lankan Spotted House Gecko is a small, approximately 4.0 in., gekkonid lizard. Background color is faded yellow-gray with dark brown markings that loosely form longitudinal rows. A dark stripe passes through the eye. The tail is banded and serrated on the sides. The venter is white. Near lights at night, individuals appear as pale pink and almost without pattern. Its center of distribution is Sri Lanka.

"Spotted House Gecko Hemidactylus parvimaculatus" by Er. Ameet Mandavia is licensed under CC BY 4.0. https://www.inaturalist.org/photos/107470033.

Introduction history and geographic range: Arrival of this species in the United States was first reported in New Orleans, Orleans Parish, Louisiana, in 2013, where it had first been collected in 2012. Introduction does not appear to have been intentional, and this gecko appears to have been introduced sometime after 2010. The Sri Lankan Spotted House Gecko is confirmed as established in New Orleans, where it inhabits buildings at the Audubon Zoo.

Ecology: The Sri Lankan Spotted House Gecko avoids trees, and frightened individuals are most apt to flee toward the ground. It is nocturnal in activity. Females lay clutches of two eggs several times each year. This species eats invertebrates and small frogs. Ecologically similar in many respects to the Mediterranean Gecko, potential for competitive interactions seems likely.

ASIAN FLAT-TAILED HOUSE GECKO
Hemidactylus platyurus (Schneider, 1792)

Identification: The Asian Flat-tailed House Gecko is a small, approximately 3.0–3.5 in., gekkonid lizard. Among adults, males are larger than females. Nondescript, light brown or gray in color with a white venter; unique in this species is the combination of the flattened serrated tail and the loose skin flap along either side of the body and rear legs. Its toes are webbed. Juveniles and adults are similar in appearance. Its center of distribution is Asia.

Introduction history and geographic range: This species is a member of the pet trade, inexpensive, and often used as a feeder lizard. Its arrival in the United States was first reported in Clearwater, Florida, in 1994, where it had been present since the 1980s. Over a ten-year period, no expansion of that colony

Photo by Suzanne Collins.

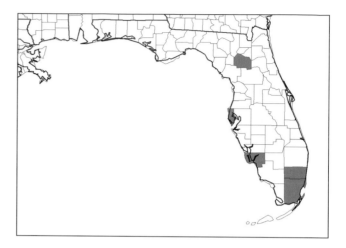

occurred from its original site. Little subsequent dispersal, human-mediated or natural, has occurred in Florida. In the United States, the Asian Flat-tailed House Gecko is found in isolated Florida localities.

Ecology: The Asian Flat-tailed House Gecko is closely associated if not exclusively with buildings and other artificial structures, where individuals can be numerous. This species is primarily nocturnal. Gravid females have been found in April, and hatchlings were found in November. This gecko is an insectivore. The Ringed Wall Gecko is a confirmed predator, and the Common House Gecko can completely displace this species from buildings. What little is known of the Asian Flat-tailed House Gecko in Florida may remain so in light of its poor ability to persist in the presence of other competitive and predatory geckos.

Mediterranean Gecko

Hemidactylus turcicus (Linnaeus, 1758)

Identification: The Mediterranean Gecko is a small, up to 4.0 in., gekkonid lizard. This species is easily identified by rows of tubercles along the back and sides of body. Its dorsum is pale in background color with brownish flecks. Its venter is white. Its tail is round in cross section, also tuberculate, and easily broken. Its center of distribution is the Middle East.

Introduction history and geographic range: Arrival of this species in the United States was through dispersal to Key West, Monroe County, Florida, via cargo ships, where it was reported in 1915 and seen there since 1910. In Alabama, it was first reported from Eufaula, Barbour County, in 1975. As reported in 1993, the Mediterranean Gecko had been seen in Mobile, Mobile County, since 1977.

Photo by Suzanne Collins.

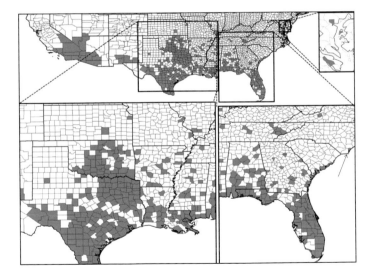

In Arizona, it was first reported from Phoenix, Maricopa County, and Tucson, Pima County, in 1973. Two Phoenix populations were established from at least 1965. Individuals were collected from the Tucson site during 1970–1972. In Arkansas, it was first reported from Fort Smith, Sebastian County, in 1990, where it had been established since the early 1970s. In California, it was first reported from Imperial County in 1988. Reported in 2005, the Mediterranean Gecko has been established on the grounds of the Moorten Botanical Gardens, Palm Springs, Riverside County, since approximately 1990. In Georgia, it was first reported from Loundes County in 1981 and reported in 1983. It was subsequently reported from Forsythe, Chatham County, in 1997, where it was collected in 1996. In Illinois, it was first reported from Marion, Williamson County, in 2006, where it had first been collected in 2002. Individuals were subsequently seen in 2011 and 2012. In Kansas, it was first reported from Lenexa, Johnson County, in 2010, where it had been collected in 2006. In Kentucky, it was first reported from Taylor Mill, Kenton County, in 2013, where it had been collected in 2012. Individuals had been seen at this site for the past seven years. In Louisiana, it was first reported from New Orleans, Orleans Parish, in 1952, where it had been present since the 1940s. In Maryland, it was first reported from Baltimore, Baltimore County, in 1991. In Mississippi, it was first reported from Oxford, Lafayette County, in 1984, where it had been collected in 1976. This colony was derived from six individuals captured in Lafayette, Louisiana, and accidentally escaped that same year. This colony did not survive the next year. The Mediterranean Gecko was again seen on the campus of Oxford University in 2007, where it was well established. Its derivation was unknown. The first established colony in Mississippi was reported from Gulfport, Harrison County, in 1993, where it had been collected in 1985. In Missouri, it was first reported from Maryland Heights, St. Louis County, in 2004, where it had been captured in 2003. The colony was still in existence as of 2006. In Nevada, it was first reported from Las Vegas, Clark County, in 1993, where it had been collected in 1992. In New Mexico, it was first found in Las Cruces, Dona Ana County, in 1992, where it had been collected in 1991. In North Carolina, it was first reported from Wilmington, New Hanover County, in 2010 and known from the site since 2003. In Oklahoma, it was known to have been introduced to a science building in Edmond, Oklahoma County, during 1962–1997, and reported in 2006. However, the first report of the species in Oklahoma was from Enid, Garfield County, in 2001. In South Carolina, it was first reported from West Ashley, Charleston County, in 2000, where it had been caught in 1998. In Tennessee, it was first reported from Memphis, Shelby County, in 2007. In Texas, it was first reported from Brownsville, Cameron County, in 1955. It was collected in 1950, and by 1954 the population was well established within several blocks downtown. In Virginia, it was first reported from Blacksburg, Montgomery County, in 2003. In the United States, the Mediterranean Gecko is found primarily in southern and coastal regions. Subsequent dispersal on its own is slow, but this species can be successful both by hard-shelled eggs and by individual geckos on vehicles and through the pet trade.

Ecology: The Mediterranean Gecko prefers stone structures and is primarily nocturnal. Unlike other exotic hemidactyline geckos established in the United States, this species avoids the brightly lighted portion of electric lights when foraging at night. Still, individuals will occupy building walls from ground to eaves. Females lay several clutches of eggs, two at a time, during warm months of the year. Consequently, its fecundity is greatest in southern populations. In parts of Florida and Texas, adults reach sexual maturity within one year of life. This small and often abundant gecko is a potential competitor with other small insectivores on buildings. In La Habra, California, this species shares retreats with the Western Fence Lizard, *Sceloporus occidentalis*. Once the dominant gecko in Florida, the Mediterranean Gecko has been replaced by other congeners, such as the Tropical House Gecko and Common House Gecko. In Texas, it is replaced by the Rough-tailed Gecko, and it remains to be seen what interactions ensue between the Mediterranean Gecko and the Sri Lankan Gecko in Louisiana. Consequently, the Mediterranean Gecko is more abundant now in states lacking its superior competitors than it is in Florida. Its tolerance for colder climate may serve to protect it in northern populations from its more tropical competitors.

Indo-Pacific Tree Gecko

Hemiphyllodactylus typus Bleeker, 1860

Identification: The Indo-Pacific Tree Gecko is a small, approximately 2.0–3.5 in., gekkonid lizard. This all-female species is easily identified by its distinctly slender body and relatively short and narrow tail. Its ground color is variable, ranging from shades of gray through brown to nearly black. At night, individuals

Photo courtesy of Fred Kraus.

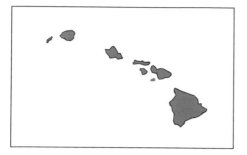

assume a very light color, in which case some of their internal organs are even more easily seen than when they assume other shades. A stripe is present from the eye to the shoulder. Its venter is light-colored, and speckling may be present. Its center of distribution is Asia.

Introduction history and geographic range: Arrival of this species in the United States probably dates to its dispersal to Hawai'i in association with the Polynesians. In the United States, the Indo-Pacific Tree Gecko is found on the Hawaiian islands of Hawai'i, Kaho'olawee, Kaua'i, Lāna'i, Maui, Moloka'i, and O'ahu.

Ecology: The Indo-Pacific Tree Gecko prefers the vegetation of forests and is primarily nocturnal. It is uncommon on buildings, where it prefers the darkest areas. The Indo-Pacific Tree Gecko usually lays two eggs under cover. It is an insectivore. Habitat modification and negative impacts by other geckos are thought to place this species at risk.

MOURNING GECKO

Lepidodactylus lugubris (Duméril and Bibron, 1836)

Identification: The Mourning Gecko is a small, approximately 3.8 in., gekkonid lizard. This all-female species can vary its ground color from light tan to shades of brown. Paired dark spots may accompany weakly defined chevrons on the dorsum. The color of the venter is cream. A stripe is present from the eye to the neck. The pattern is distinctive in hatchlings. Its center of distribution is southeastern Asia.

Introduction history and geographic range: Arrival of this species in the United States probably dates to its dispersal to Hawai'i in association with the Polynesians. In Florida, it was first reported from Port Saint Lucie, St. Lucie County, in 2011, from specimens collected in 2005. This colony, highly localized, was derived through the pet trade. In the United States, the Mourning Gecko is found on the Hawaiian islands of Hawai'i, Kaho'olawee, Kaua'i, Lāna'i, Maui, Moloka'i, Ni'ihau, and O'ahu, and in southern Florida.

"Lepidodactylus lugubris" by dhobern is licensed under CC BY 2.0. https://www.flickr.com/photos/25401497@N02/13399422443.

Ecology: The Mourning Gecko can excel in the presence of humans, capable of living well both inside and outside inhabited buildings. Under ideal situations, population sizes can be large. Pseudo-copulation with a conspecific stimulates production of one or two eggs laid under cover. In captivity, clutches can be produced every month or so, hatch in two months, and sexual maturity is reached before first birthday. The Mourning Gecko will perch in lighted areas to consume light-attracted insects. Its diet also includes nectar. In Hawai'i, the once abundant Mourning Gecko was negatively impacted by the Common House Gecko, primarily though competition. Arrival of the Red-vented Bulbul, *Pycnonotus cafer,* in the late 1900s has had a negative impact on Mourning Gecko through predation. The Common House Gecko, a superior competitor of the Mourning Gecko, also occurs in Florida, as does the Tropical House Gecko. A prediction can be made of its demise in St. Lucie County after the inevitable colonization by the presently widespread Tropical House Gecko.

MADAGASCAN GIANT DAY GECKO
Phelsuma grandis Gray, 1870

Identification: The Madagascan Giant Day Gecko is a large, up to 9.0–11.0 in., gekkonid lizard. Among adults, males are larger than females. This gecko is bright green above. A red stripe runs from the nostril to the eye. Often, a small red spot is behind each eye. Orange to red spots of varying size and number are on the dorsum, especially from the lower half to the rear legs. The venter is uniformly whitish in color. Individuals under stress are a dull green. Its center of distribution is Madagascar.

Introduction history and geographic range: Arrival of this species in the United States was through the pet trade, where it was reported as breeding in residential areas of Broward County, Florida, and seen in Ft. Myers, Lee County, Florida, in 1999. Establishment of this species in Florida was confirmed following its detection on several of the Florida Keys, Monroe County, in 2003 from 2002

"Grosser Madagaskar-Taggecko" by nyffy is marked with CC0 1.0. https://www.flickr.com/photos/64872657@N08/49320841112.

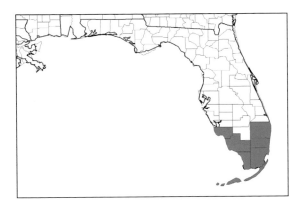

collections. Intentional human-mediated dispersal has resulted in rapid colonization by this species in southern mainland Florida and the Florida Keys. In the United States, the Madagascan Giant Day Gecko is found in southern Florida.

Ecology: The Madagascan Giant Day Gecko prefers large trees for its arboreal existence but will also display and hunt from the sides of buildings. It is primarily diurnal, and males are strongly territorial. Consequently, the Madagascan Giant Day Gecko can initially be mistaken for one of the larger anoles in southern Florida. Captive females are highly fecund, producing two-egg clutches of adhesive eggs every two to four weeks throughout the year. Its reproductive cycle in southern Florida is unknown, although neonates were found on the Florida Keys in October. Captives can also reach sexual maturity before the end of their first year of life. Its diet comprises invertebrates, nectar, sweet exudate from fruits, and smaller lizards. It is a predator of the Northern Curly-tailed Lizard. In some instances, individuals will hunt from lights at night, placing them in direct contact with nocturnal geckos. Primarily diurnal in activity, the Madagascan Giant Day Gecko certainly encounters anoles. Interspecific predation in either direction or competition for perch sites, or both, are potential outcomes of interactions that could ultimately shape those lizard assemblage structures.

ORANGE-SPOTTED DAY GECKO

Phelsuma guimbeaui Mertens, 1963

Identification: The Orange-spotted Day Gecko is a medium-sized, approximately 7.0 in., gekkonid lizard. Among adults, males are larger than females. The ground color is green to bright green and is interrupted by irregular orange lines and spots. Its nape and tail tip are blue. The venter is light yellow. Dorsum color of juveniles is brown or gray and lined with rows of small light-colored spots. Its center of distribution is Mauritius.

Photo by Suzanne Collins.

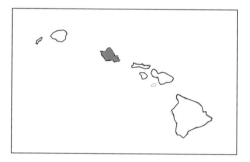

Introduction history and geographic range: Arrival of this species in the United States was thought to be associated with the pet trade, where it was first seen on Oʻahu, Hawaiʻi, in the mid-1980s. It was reported in 1996. In the United States, the Orange-spotted Day Gecko is found on the Hawaiian island of Oʻahu.

Ecology: The Orange-spotted Day Gecko is found in residential neighborhoods in association with large trees. It is diurnal in activity. Multiple two-egg clutches are laid during at least spring–fall. Eggs are adhesive and hard-shelled and may be placed with those of other females of the same species in tree cavities. The Orange-spotted Day Gecko feeds on insects, nectar, and soft fruit. Ecological impacts on or by this species are unknown in Hawaiʻi.

Gold Dust Day Gecko
Phelsuma laticauda (Boettger, 1880)

Identification: The Gold Dust Day Gecko is a small, approximately 5.5 in., gekkonid lizard. The ground color is green to bright green. As per its common name, the yellow markings behind the neck to just beyond its front legs resemble a generous sprinkling of gold dust. Shades of blue are found on the legs and feet and around the eyes. Reddish bars are present on the head. Three dull to bright red elongate markings begin midway along the dorsum and break into small spots at a point near the rear legs. Its venter is uniformly dull white. Its center of distribution is Madagascar and the Comoros.

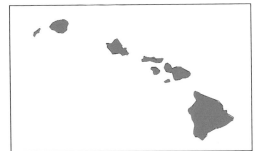

"Gold dust day gecko (Phelsuma laticauda)" by Waifer X is licensed under CC BY 2.0. https://www.flickr.com/photos/47327870@N00/16040401796.

Introduction history and geographic range: Arrival of this species in the United States was associated with the release of pets in the upper Manoa Valley on Oʻahu, Hawaiʻi, in 1974 and reported in 1996. In the United States, the Gold Dust Day Gecko is found on the Hawaiian islands of Hawaiʻi, Kauaʻi, Maui, and Oʻahu.

Ecology: The Gold Dust Day Gecko is found in residential neighborhoods, where it lives on vegetation and buildings. The Gold Dust Day Gecko is primarily diurnal in activity. Multiple two-egg clutches are laid at intervals of two to four weeks during the warmer months. Eggs are adhesive and hard-shelled and can hatch as soon as just over one month. Sexual maturity is reached by one year of life. The Gold Dust Day Gecko feeds on insects, nectar, and soft fruit. It will also eat smaller lizards. Some individuals will hunt near lights at night for insects, placing them in direct contact with nocturnal geckos. Ecological impacts on or by this species are unknown in Hawaiʻi.

RINGED WALL GECKO

Tarentola annularis (Geoffrey Saint-Hilaire, 1827)

Identification: The Ringed Wall Gecko is a small, up to 6.0 in., gekkonid lizard. Among adults, males are larger than females. The body is stocky in build, and the dorsum and sides are patterned in earth tones of beige, tan, and light brown. Two pairs of white spots are present on the shoulders. Its venter is uniformly off-white in color. The tail is laterally serrated in shape. The center of distribution of the Ringed Wall Gecko is northern Africa.

Photo courtesy of Jake Scott.

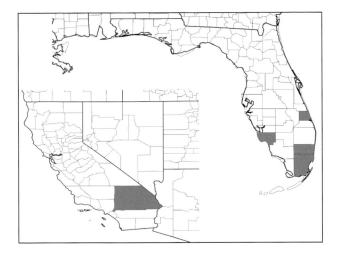

Introduction history and geographic range: Arrival of this species in the United States was through the pet trade, where it was reported from Miami-Dade and Lee counties, Florida, in 1997, where it had been present since 1990. A colony was also confirmed as established in Redlands, San Bernardino County, California, in 2014, where it had been present since 1995 or 2002. In the United States, the Ringed Wall Gecko is established in southern Florida. It is also well established locally in California.

Ecology: All colonies of the Ringed Wall Gecko are associated with buildings. It is nocturnal and active throughout the year. In Homestead, up to five two-egg clutches are laid annually. It feeds on invertebrates and lizards, and in Homestead, Florida, individuals were found to have eaten Asian Flat-tailed House Geckos and Common House Geckos. It is likely a predator of any of the nocturnal geckos it may encounter. It seems likely that the Tokay Gecko and Cuban Treefrog are potential predators of this species.

Moorish Gecko

Tarentola mauritanica (Linnaeus, 1758)

Identification: The Moorish Gecko is a small, up to 6.0 in., gekkonid lizard. A stocky build, gray-brown background color, and large, almost pointed dorsal scales are the reasons for its pet trade name, the Crocodile Gecko. Adults are drab in color and may be strongly or weakly banded dorsally. Banding is evident on the tail. The venter is yellowish or white. Hatchlings have well-defined transverse bands. Its center of distribution is northern Africa.

Introduction history and geographic range: Arrival of this species in the United States was through the pet trade, where it was reported from Ft. Myers,

"Moorish Gecko (Tarentola mauritanica) (found by Jean NICOLAS)" by Bernard Dupont is licensed under CC BY-SA 2.0. https://www.flickr.com/photos/berniedup/34844955103.

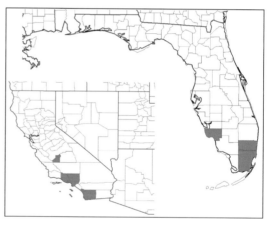

Lee County, and Miami, Miami-Dade County, Florida, in 1999. It had been present since at least 1996. In California, it was first reported from El Cajon, San Diego County, in 2010, where it was first captured in 1997. The population was well established and derived from a local pet shop that sold exotic reptiles. In the United States, the Moorish Gecko is found in isolated colonies in southern Florida and southern California.

Ecology: The Moorish Gecko is strongly associated with buildings. It is nocturnal and active throughout the year. Clutches of one or two eggs are produced. A neonate was captured in Broward County, Florida, in May. This species feeds on invertebrates and presumably lizards and is likely a predator of any of the nocturnal geckos it may encounter. It seems likely that the Tokay Gecko and Cuban Treefrog are potential predators of this species.

Sphaerodactyl Geckos: Sphaerodactylidae

YELLOW-HEADED GECKO

Gonatodes albogularis (Duméril and Bibron, 1836)

Identification: The Yellow-headed Gecko is a small, up to 3.25 in., sphaerodactylid lizard. This species is strongly sexually dimorphic in color. Among adults, males have a dark body with a blue tint. The head is yellow, and a dark shoulder spot is present. The tail has a white tip. Females are light brown in color. Toe pads are not present on this gecko. Its center of distribution is the West Indies.

Introduction history and geographic range: Populations of this species in the United States were first reported from Key West, Monroe County, Florida, in 1939, and thought to have been there for many years. Commerce appears to have been the agent of its dispersal to Florida. In the United States, the Yellow-headed Gecko is established on Key West, Florida.

Photo by Suzanne Collins.

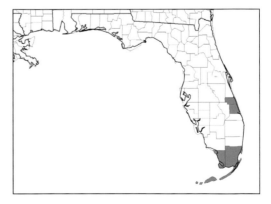

Ecology: In southern Florida, the Yellow-headed Gecko prefers shady areas of buildings and the trunks of large trees and is diurnal in activity. A clutch generally comprises one non-adhesive egg. Incubation lasts approximately two months. This species is an insectivore. At one time, this gecko was also found on other Florida Keys (Bahia Honda, Boca Chica, Key Largo, Stock Island) and in Miami-Dade, Broward, and St. Lucie counties. This small gecko was at one time abundant. Since the 1980s, the Yellow-headed Gecko has actually become rare in Florida. The establishment of more aggressive hemidactyline geckos was thought to be a cause of its decline.

OCELLATED GECKO
Sphaerodactylus argus Gosse, 1850

Identification: The Ocellated Gecko is a small, up to 2.5 in., sphaerodactylid lizard. This gecko is recognized by a dark striated pattern over a flesh-colored to orange dorsal background. The tail may be especially orange to red in color. Light-colored ocelli are present from the head through the shoulders. The center of its geographic distribution is the Caribbean.

Introduction history and geographic range: Its arrival in the United States is thought to have been incidental in association with commerce. The Ocellated Gecko was first reported in the United States on Key West, Monroe County, Florida, in 1954. In the United States, the Ocellated Gecko is established on Key West and Stock Island, Monroe County, Florida.

Ecology: The Ocellated Gecko can be found in arboreal situations but is most often encountered under terrestrial cover. It is primarily nocturnal in activity.

Photo courtesy of Michael R. Rochford.

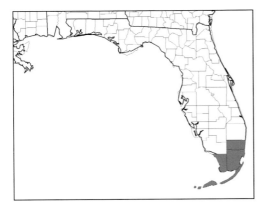

Single-egg clutches are laid; however, the duration of nesting season and number of clutches produced each year are unknown for this population. This species feeds on small invertebrates. The Ocellated Gecko is actually rare in Florida, with a tenuous hold on persistence in the Florida Keys. Never known to be abundant in Florida, the Ocellated Gecko was at one time even thought to have been extirpated. The reasons for its actual rarity are unknown, although competition with other *Sphaerodactylus* species and juveniles of *Hemidactylus* species cannot be ruled out as potential causes. Likewise, predation by larger geckos and by the Cuban Treefrog, cannot be ruled out as contributing to its precarious existence in Florida.

ASHY GECKO

Sphaerodactylus elegans MacCleay, 1834

Identification: The Ashy Gecko is a small, up to 2.8 in., sphaerodactylid lizard. Among adults, males are larger than females. The Ashy Gecko is tan to light brown in ground color and heavily stippled in brown throughout. The tail, also stippled in brown, is light orange in ground color. Juveniles could easily be mistaken for a different species. Their gray head, yellow-orange body, and bright orange tail are crosswise banded in black. The center of its geographic distribution is the West Indies.

Introduction history and geographic range: Its arrival in the United States is thought to have been incidental in association with commerce. The Ashy Gecko was first detected in the United States on Key West, Monroe County, Florida, in 1922. In the United States, the Ashy Gecko is found in Broward, Miami-Dade, and Monroe (Florida Keys) counties, Florida.

Ecology: The Ashy Gecko can be found on buildings and trees at night. Individuals can also be found under moist ground debris and rotting logs, and they can be uncovered from bark by day or night. Single-egg clutches are laid. Ovigerous

"Sphaerodactylus elegans.jpg" by thibaudaronson is licensed under CC BY-SA 4.0.
https://commons.wikimedia.org/w/index.php?curid=88815504.

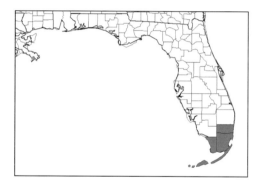

females have been found in March and September, and females will communally nest with conspecifics and with other species of geckos. Small invertebrates comprise its diet. The Ashy Gecko is likely prey of larger geckos and by
the Cuban Treefrog. Once abundant on the lower Florida Keys, the Ashy Gecko
was subject to rapid replacement by the Tropical House Gecko. Potential for
a competitive relationship with the rare Ocellated Gecko remains unstudied.

Iguanas: Iguanidae

MEXICAN SPINY-TAILED IGUANA
Ctenosaura pectinata (Wiegmann, 1834)

Identification: The Mexican Spiny-tailed Iguana is a large, up to 48.0 in., iguanid lizard. Among adults, males are larger than females. Adults are black with white or yellow blotches. Females may have green-hued peach markings. The spiny dorsal crest is more pronounced in the male. Juveniles are bright green with scattered black markings. Similar in appearance to the Gray's American Spiny-tailed Iguana, this species can be identified by having three rows of flat

"Ctenosaura pectinata" by Dick Culbert is licensed under CC BY 2.0. https://www.flickr.com/photos/92252798@N07/24333259842.

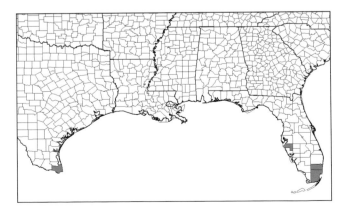

scales on the tail between each of the first five or six whorls of large spiny scales. Subsequently, two rows of flat scales are between each of the next five or six whorls of large scales. Its center of distribution is Mexico.

Introduction history and geographic range: Arrival of this species in the United States was through the pet trade, where it was reported from Miami, Miami-Dade County, Florida, in 1978. It was also reported from Brownsville, Cameron County, Texas, in 2008. In the United States, the Mexican Spiny-tailed Iguana is found in localized areas of Broward and Miami-Dade counties, Florida. It is also well established locally on the grounds of the Gladys Porter Zoo in Brownsville, Texas.

Ecology: The Mexican Spiny-tailed Iguana excels around stone structures, such as rock walls. It readily climbs trees, and in southern Florida, it is in part associated with disturbed pineland and hammock and associated residences. It is diurnal and most active in hot weather. Individuals are very wary, and very hard to catch without the use of a trap. Capture by hand is not to be taken lightly. Males are territorial. Gravid females have been found in June in southern Florida. This lizard is primarily an herbivore and frugivore, although it will opportunistically eat small vertebrates. This species has remained static in its Miami-Dade County distribution and occurs in only one other area of Florida. Its colonization pattern stands in sharp contrast to that of its congener, Gray's American Spiny-tailed Iguana, whose introduced range subsumes that of the Mexican Spiny-tailed Iguana and whose population sizes can be large. The reasons for this difference in colonization success are unknown.

GRAY'S AMERICAN SPINY-TAILED IGUANA
Ctenosaura similis (Gray, 1831)

Identification: The Gray's American Spiny-tailed Iguana is a large, up to 55.0 in., iguanid lizard. Among adults, males are larger than females. Adults are typically some shade of gray in background color. Black dorsal crossbands reach the venter. A spiny dorsal crest is present and more developed in the male. During the breeding season, adults may develop an orange hue. Juveniles are bright green with black, brown, and white markings. Similar in appearance to the Mexican Spiny-tailed Iguana, this species can be identified by having two or three rows of flat scales on the tail between each of the first, second, and sometimes third whorls of large spiny scales. Subsequently, two rows of flat scales are between each of the next six to eight whorls of large scales. Its center of distribution is Central America. Its native geographic range is larger than that of any of the other *Ctenosaura* species.

Introduction history and geographic range: Arrival of this species in the United States was through the pet trade, where it was first reported from Miami,

Photo courtesy of Janson Jones.

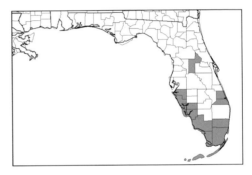

Miami-Dade County, Florida, in 1999. However, in 1979, three individuals were released onto Gasparilla Island, Charlotte and Lee counties, and by 2010, population size numbered in the tens of thousands. In the United States, the Gray's Spiny-tailed Iguana is established in southern Florida.

Ecology: Gray's Spiny-tailed Iguana thrives in open sandy habitat with some vertical structure. It is an excellent climber and sleeps and escapes into burrows of its own making or those of other animals, such as the Gopher Tortoise. Diurnally active, individuals are most conspicuous during the heat of the day. They are extremely wary and difficult to catch without the use of a trap. Capture by hand is not to be taken lightly. Males are territorial. Clutch sizes average 32 eggs and increase in number with increasing size of the female. Eggs are laid in open sandy sites, and hatchlings appear in late summer through early fall. Growth rates of Florida populations of this species are slower than those within their native ranges. In southern Florida, diet shifts ontogenetically from primarily insects when juvenile to primarily vegetation when adult. Included in its diet are berries of Brazilian pepper and flowers of beach sunflower. It will opportunistically eat small vertebrates. It is depredated by the Bobcat, *Lynx rufus,* and Raccoon.

The removal of the overabundant Raccoon at a state park in Miami-Dade County resulted in a rapid increase in numbers of this previously uncommon lizard. Unlike its congener, this species is comparatively very successful, having expanded its introduced range somewhat continuously and in large populations. To that end, 12–15 burrows per 100 m linear habitat bordering a Brazilian pepper stand were recorded on Gasparilla Island. The plant provided food and cover. Colonization success by Gray's American Spiny-tailed Iguana presents several threats to the integrity of the systems in which it is established. Use of Gopher Tortoise burrows was shown to inhibit use by Gopher Tortoises when occupied by multiple individuals. Moreover, a hatchling Gopher Tortoise was recovered from a stomach in a sample of lizards examined for diet. Other burrowing species, such as the Florida Burrowing Owl, *Athene cunicularia floridana,* as well as some of the commensal species associated with Gopher Tortoise burrows, could be negatively impacted by this lizard by its presence or depredations. In one instance, a Southern Black Racer was attacked and killed by a Gray's Spiny-tailed Iguana, both approximately 24.0 in. The snake was not eaten, and the carcass was still there the next day. Ground-nesting birds could also be at risk from disturbance or predation of nests. Its active foraging on Brazilian pepper berries could serve to disperse its seeds in scat, thereby benefiting both species.

The extent to which property damage was associated with the high numbers of Gray's American Spiny-tailed Iguana on Gasparilla Island prompted residents of both counties on the island to approve a self-tax to support intensive control efforts. During 2008–2011, 15,000 Gray's American Spiny-tailed Iguanas were removed from the island; nearly 3,000 were necropsied for life history study. The control program is ongoing to inhibit a second population explosion. One measurable effect of its removal was an increase in hatchling Gopher Tortoises.

GREEN IGUANA
Iguana iguana (Linnaeus, 1758)

Identification: The Green Iguana is a large, up to 79.0 in., iguanid lizard. Among adults, males are larger than females. The exotic population of the Green Iguana in Florida comprises a melting pot of many geographically distinct forms derived from a wide geographic range. Adult males are various shades of green, tan, or bronze. The body may or may not have black transverse lateral bands. A spiny dorsal crest is present, but the length of the spines varies to up to 3.0 in. in height. The tail is smooth and is usually ringed in black. The head is typically,

Photo courtesy of Carlos Nieves.

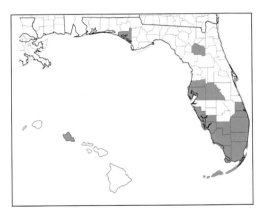

but not always gray, and jowls develop with age. At least the face, shoulders, and front legs turn orange during the short winter breeding season. This color change often extends partway on the body, and occasionally throughout the body, and includes the rear legs. Likewise, the intensity of the orange color varies, with some males turning to a nearly pumpkin-orange hue throughout. Adult females are generally dull olive or brown in ground color, and the tail is usually ringed in black. Their spiny dorsal crest is usually much reduced in height compared with that of the male. Still, there are exceptions. Hatchlings and young adults are usually a vibrant green. Dewlaps are present in both sexes. Occasionally, individuals will have hornlike projections between the nostrils. These projections generally range from one to three in number and to 0.5 in. in height. Its center of distribution is Central and South America.

Introduction history and geographic range: Arrival of this species in the United States was through the pet trade, where it was reported from Oʻahu, Hawaiʻi, in the 1950s, and, on the mainland, in Miami, Miami-Dade County, Florida, in 1966. In the United States, the Green Iguana is found throughout much of southern Florida. Its status on the Hawaiian island of Oʻahu is uncertain.

Ecology: Ideal habitat for the Green Iguana is secondary growth with large trees alongside or nearby lentic or slow-moving bodies of fresh and brackish water. It will forage and bask on the ground and on branches of suitable diameter. The Green Iguana is diurnally active. Sleeping individuals are most often encountered well above the ground, but they also use burrows of their making or those of others, such as the Florida Burrowing Owl. Despite their large body size, they can remain easily undetected, even when basking in a group. Population sizes can be large. For instance, 397 individuals were removed from a South Florida site in five years. The population was estimated to be 241.7 individuals/mi² following the removal of 263 Raccoons from the site. Elimination of an overabundant Raccoon population elsewhere in southern Florida likewise resulted in a population explosion of Green Iguanas. An estimated 4,433 burrows/acre was calculated along a canal levee in southern Florida. More than 20 individuals/linear mi. were counted foraging or basking along a canal in southern Florida. Below 60°F, movement is noticeably impaired. Episodic frosts in southern Florida are accompanied by Green Iguana carcasses having dropped from arboreal perches or found in situ where they died while basking. Capture by hand is not to be taken lightly. Sharp claws, a whiplike tail, and a very strong bite become a dangerous combination. Frightened individuals will run or readily dive into water, remaining underwater for some time.

Males are highly territorial. Females may share the territory of a male for an extended time or may wander among territories. Mating occurs for a short time during winter–spring. During the mating season, males are conspicuous in color and habits. Females lay between 17 and 49 eggs during March–July, with a peak in nesting in April. Nesting sites are nearly always in sandy soil and in the open. Hatchlings appear during June–September, and sexual maturity is reached earlier in males (16–17 mo.) than females (24–25 mo.).

Diet in Florida is overwhelmingly herbivorous. Leaves, flowers, fruits, and grasses are typical fare. Exceptionally, scavenging and ingestion of invertebrates is noted as well. When abundant, this species can negatively impact a food source, one example being the abandonment of a *Hibiscus* garden that could no longer be maintained at a botanical garden due to extensive browsing by Green Iguanas. The Yellow-crowned Night Heron, *Nyctanassa violacea*, and the threatened Florida Burrowing Owl eat young Green Iguanas. Eggs are eaten by the Gray Fox, *Urocyon cinereoargentus,* and Raccoon. Young and adults are subject to the depredations of domestic dogs and Raccoons, and along canals, the Burmese Python is a potential predator.

Tracks of adult Green Iguanas have been seen in the direction of entering burrows of the Florida Burrowing Owl, which through disturbance alone could negatively impact the nests of this state-listed threatened species. Burrows of the Gopher Tortoise are altered by large adults. Population explosions of the Green Iguana are an unintended consequence of removal of superabundant populations of the Raccoon, a midlevel carnivore with few predators of its own in urban areas. Dispersal of Brazilian Pepper through ingestion of berries is a consequence of its establishment. Inclusion of Grey Nicker-Bean, *Caesalpinia bonduc,* in its diet is detrimental to the Miami Blue Butterfly, *Cyclargus thomasi,* whose eggs are laid on that species.

Quantified economic costs of negative impacts by the Green Iguana have been calculated with respect to destruction of infrastructure damage from burrowing, and individual Green Iguanas are documented airstrike hazards on airport runways. In Puerto Rico, incidence of damage was high enough to result in costly interference with airport operations and to necessitate control efforts. This same damage has been identified as actual and potential threats at various airfields in South Florida, including Homestead Air Reserve Base and various airports in Miami, Ft. Lauderdale, and West Palm Beach among others. Single engine military jets are most at risk.

Wall Lizards and Lacertas: Lacertidae

WESTERN GREEN LACERTA
Lacerta bilineata Daudin, 1802

Identification: The Western Green Lacerta is a medium-sized, up to 16.0 in., lacertid lizard. Adults are green above and yellow underneath. Among adults, males are usually brighter in color than females. The face and throat of the male is bright blue in color. Juveniles are brownish above and yellow underneath. Two to four longitudinal stripes may be present. Its center of distribution is western Europe.

Introduction history and geographic range: Arrival of this species in the United States was through the pet trade, where it was reported from Topeka, Shawnee County, Kansas, in 1974. The population, established since prior to 1968, was derived from escaped animals from a pet dealer. A 19-year hiatus in captures since 1977 led to the conclusion that the population was extirpated. Subsequent captures and observations confirmed its persistence in Topeka. In the United States, the Western Green Lacerta is found in Topeka, Kansas.

Ecology: In Topeka, the Western Green Lacerta is found both in trees and on the ground. This species is ecologically akin to the large branch anoles found in Florida and Hawai'i. The Western Green Lacerta is diurnal and, in Kansas it

Photo by Suzanne Collins.

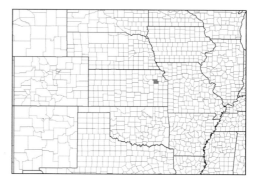

is active from late April to mid-October, with peak activity between mid-May and early September. Males are territorial. Mating occurs in May and early June. Eggs are laid in the summer. Primarily an insectivore, it eats other lizards, including the Italian Wall Lizard.

COMMON WALL LIZARD
Podarcis muralis (Laurenti, 1768)

Identification: The Common Wall Lizard is a small, up to 8.0 in., lacertid lizard. Individuals are highly variable in color. Most commonly, the head and dorsum are brown with spots or with a reticulated or speckled pattern. Sides are darker brown, and light bars may be present. The venter is subject to polymorphism.

Blue spots are present above the hips of adult males. Its center of distribution is Europe.

Introduction history and geographic range: Arrival of this species in the United States was through a deliberate release of ten individuals in Cincinnati, Hamilton County, Ohio, that were transported from Lake Garda in northern Italy in 1951. This colony was reported in 1977. In Indiana, it was first reported from Clarksville, Clark County, in 2005, where it was collected in 2005. In Kentucky, it was first reported from Park Hills, Kenton County, in 2002, where it was collected in the summer of 1998. In the United States, the Common Wall Lizard occurs in a tri-state region of Ohio, Indiana, and Kentucky. The southern Indiana population may or may not have been founded naturally, whereas the northern Kentucky population is thought to be the result of intentional introduction from Cincinnati.

Photo by Suzanne Collins.

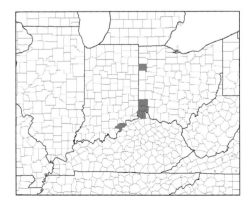

Ecology: Debris piles and unmortared stone retaining walls in urban areas are ideal habitats for the Common Wall Lizard. Railroads can be suitable linear habitat, thereby providing dispersal pathways. Habitat must be sunny. It is diurnal in activity. In Cincinnati, population sizes can range 1,114–3,100 lizards/acre, with home range sizes of up to 15 males/m² and 10 females/m². The Common Wall Lizard is highly fecund, with females producing 4–5 egg clutches up to three times each year. It is an insectivore. Feral cats are predators of this lizard. Where the Common Wall Lizard and the Italian Wall Lizard co-occur, the latter is displaced from favored structures. Potential for high population sizes, high fecundity, and ability to take advantage of railroad corridors and, possibly, by rafting are advantages to its colonization success and future geographic range expansion. On the other hand, clean-up of city debris and replacement of unmortared walls with those of concrete reduce availability of suitable refuges and result in population declines.

ITALIAN WALL LIZARD

Podarcis siculus (Rafinesque, 1810)

Identification: The Italian Wall Lizard is a small, up to 8.0 in., lacertid lizard. Among adults, males are larger than females. The dorsum is bright to dull green, often accompanied by a mid-dorsal stripe or spots. Blue spots may be present on the shoulder. Sides are mottled in brown and white. Venter is white or gray. Its center of distribution is Europe.

Introduction history and geographic range: Arrival of this species in the United States was through the pet trade, where it was reported from Philadelphia,

Photo by Suzanne Collins.

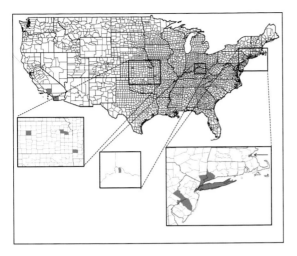

Philadelphia County, Pennsylvania, in the 1920s. The Philadelphia population may be extinct. Individuals were found at a site in Bucks County, Pennsylvania, in 2017 that were derived from intentional releases in 2014 and 2015. As of 2017, this species was considered established at the site. In California, it was first reported from San Pedro, Los Angeles County, in 2010. The population was derived from seven adults from Taormina, Sicily, in 1994. In Connecticut, it was first reported from Greenwich, Fairfield County, in 2014. In Kansas, it was first reported from Topeka, Shawnee County, in the 1950s. This species is also established in Hays in Ellis County, Lawrence in Douglas County, and Topeka, where populations are large but localized. Kansas populations were derived from Tuscany. In Massachusetts, it was first reported from Fenway Victory Gardens, Boston, Suffolk County, in 2017. In Missouri, it was first reported from Joplin, Jasper County, in 2015, where the population had been in existence since 2001. The source of the Joplin population comprised nine lizards from Topeka, Kansas, that had escaped the same year. In New Jersey, it was first reported from Mt. Laurel, Burlington County, in 2010, presumed to have been intentionally introduced in 1984. The New Jersey population appears to have been derived from the Adriatic coast. In New York, it was first reported from Garden City in 1975, where individuals had been released from a pet shop in 1967. New York populations were derived from Tuscany.

Ecology: In the United States, the Italian Wall Lizard is found in increasing locations of northern latitudes from coast to coast. The Italian Wall lizard is most at home above the ground on rock piles and on walls, particularly those made of stone. For this reason, cities are especially likely to support this species. Activity is diurnal, and, in New York, individuals are active during April–October. One or two clutches of two to six eggs are laid. The Italian Wall Lizard is primarily an insectivore but will also include vegetation in its diet. On Long Island, a wide range of taxa was eaten. Prey diversity and evenness were lower in females than males. Males ate fewer items and more hard-bodied prey than did females. Feral cats are predators, and a Common Gartersnake was observed to hunt them. Individuals can supercool if kept dry. Sufficiently deep hibernacula of at least 9.5 in. are a limiting factor in their northward expansion. This species is actively displaced from favored perch sites by the Common Wall Lizard.

Curly-tailed Lizards: Leiocephalidae

NORTHERN CURLY-TAILED LIZARD

Leiocephalus carinatus Gray, 1827

Identification: The Northern Curly-tailed Lizard is a large, up to 8.5 in., leio-cephalid lizard. Among adults, males are larger than females. This lizard is of stocky build with heavily keeled scales on a brown or gray dorsum. The neck is nearly as thick as the base of the skull, adding to its stocky appearance. Rows of small black dorsal spots are present. Dorsal tail color of adult males is the same as the color of the body. The tail is banded in females and juveniles. True to its common name, tails are usually held high in an upward curl, in contrast to the Red-sided Curly-tailed Lizard whose tail curling is much less developed. Its center of distribution is the West Indies.

Introduction history and geographic range: Arrival of this species in the United States was associated with pest control, where it was reported from Palm Beach

Photo courtesy of Gary Busch.

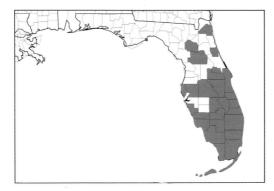

County, Florida, in 1958. The colony may have been established since the 1940s. Subsequent introductions were through the pet trade. In the United States, the Northern Curly-tailed Lizard occurs throughout much of southern Florida, where it can be very abundant along the east coast.

Ecology: The Northern Curly-tailed Lizard is nearly exclusively terrestrial in habits, although uncommonly, individuals will climb trees. Open habitat with rocks is required for this diurnally active species to thrive. In Florida, its habitat requirements are often met in the form of rock walls, including along sidewalks and tennis courts, with access to a refuge below the surface. These sorts of habitats are often inhabited by the Brown Anole, which can be an abundant and nutritious source of food for the Northern Curly-tailed Lizard. It is active throughout the year. In warmer weather, daily activity occurs in two periods, whereas in cooler weather, individuals are active from late morning continuously through the afternoon. Females lay a single clutch averaging four eggs between May and August; however, a second clutch, although rare, is possible. Sexual maturity is reached within six months of hatching. This species preys on invertebrates, including crickets, grasshoppers, beetles, roaches, and small vertebrates. The Brown Anole is subject to predation by the Northern Curly-tailed Lizard. A variety of bird species prey on the Northern Curly-tailed Lizard, including the Northern Mockingbird, Loggerhead Shrike, *Lanius ludovcianus,* and Little Blue Heron, among others. The Madagascan Giant Day Gecko preys on this species. This species poses a threat to sandy upland species, such as the Eastern Six-lined Racerunner, *Aspidoscelis sexlineata sexlineata,* and the Florida Scrub Lizard, *Sceloporus woodi,* that can exist even marginally along the coast. Potential competition for food may exist between the Northern Curly-tailed Lizard and the Brown Anole.

Red-sided Curly-tailed Lizard

Leiocephalus schreibersii (Gravenhorst, 1837)

Identification: The Red-sided Curly-tailed Lizard is a large, up to 10.0 in., leioce-phalid lizard. Among adults, males are larger than females. Adding to its stocky appearance, the neck is nearly as thick as the base of the skull. This lizard is of stocky build and has a brownish-gray dorsum. Red bands run vertically along the sides of the body with blue patches between them. Fore and rear legs may have turquoise highlights. The tail is banded and flecked in red underneath. A lateral fold is present. Females and juveniles are paler in color than adult males. This species does not hold its tail curled as prominently as the Northern Curly-tailed Lizard. Its center of distribution is the West Indies.

Introduction history and geographic range: Arrival of this species in the United States was associated with the pet trade, where it was reported from Miami-Dade County, Florida, in 1983. The colony had been in existence since 1978. Extirpated in 1981, individuals of that colony gave rise to a second colony in Miami Lakes, Miami-Dade County, Florida. In the United States, the Red-sided Curly-tailed Lizard occurs in isolated populations in southern Florida.

Photo by Suzanne Collins.

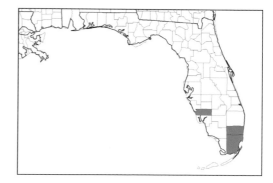

Ecology: This species is nearly exclusively terrestrial in habits, although uncommonly, individuals will climb trees. The Red-sided Curly-tailed Lizard is less dependent on open rocky habitat than its more successful close relative, the Northern Curly-tailed Lizard. It is diurnal in activity and active throughout the year. The nesting season has not been identified in southern Florida; however, hatchlings have been found in June. This species is an insectivore, although inclusion of small vertebrates in its diet seems likely. Very little is known about the biology of this species in Florida. Independently introduced populations of this species may provide an advantage to what had historically been a generally unsuccessful colonizing species in Florida.

Horned Lizards: Phrynosomatidae

Texas Horned Lizard
Phrynosoma cornutum (Harlan, 1825)

Identification: The Texas Horned Lizard is a large, up to 5.0 in., phrynosomatid lizard. Adult size of both sexes varies with latitude, with the largest from southern locations. Among adults, males are smaller than females. Squat and round in shape, horned lizards of any species are colloquially referred to as horned toads. Background color is tan to light brown and overlaid with some elongate dark brown mottling on the face and sides. A light-colored vertebral stripe is present. The venter is white, and the tail is very short. Its most distinguishing feature is the very well-developed spines behind the head, spines on the dorsum, and the small but numerous spines that form two rows along the sides.

Photo by Suzanne Collins.

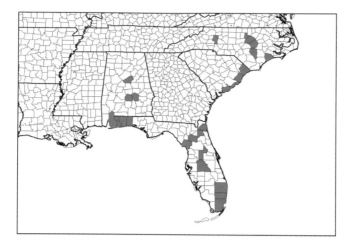

Handled, this lizard has a rough sandpaper-like feel to it. Its center of distribution is the American Southwest.

Introduction history and geographic range: This species was first detected in Raleigh, Wake County, North Carolina, in 1880. In Florida, it was first reported in Palm Beach, Palm Beach County, in 1928, and in Miami, Miami-Dade County, in 1934. In Alabama, it is established in the central part of the state. In Florida, it is well established in northern parts of the state; the Duval County population has been known since the early 1950s. In North Carolina, a well-established population was detected in Onslow County in the late 1980s and thought to have been present for approximately ten years. In South Carolina, populations are established on two barrier islands in Charleston County and on beachfront of a portion of Georgetown and Horry counties.

Ecology: The Texas Horned Lizard is a creature of open sandy desert and grassland. Consequently, its success in extralimital natural and disturbed sandy coastal areas is not surprising. The species is diurnal and most active during summer. Clutch size is large and varies in size latitudinally. Eggs are laid during the warmer months. It is well adapted in its morphology to the life of a sit-and-wait predator. It specializes in eating ants; however, little is known of its ecology in introduced areas. West of the Mississippi, the Texas Horned Lizard encounters several monomorphic species of harvester ants (*Pogonomyrmex* spp.) on which it feeds. Only one harvester ant species, the Florida Harvester Ant, *P. badius,* is found east of the Mississippi. This ant species is polymorphic and is restricted to sandy soils of the Coastal Plain states no farther north than North Carolina. The impact of the Texas Horned Lizard on the Florida Harvester Ant is unknown, as is the extent to which this ant prey species might be a limiting factor in the introduced range of its predator.

Eugongylid Skinks: Eugongylidae

PACIFIC SNAKE-EYED SKINK

Cryptoblepharus poecilopleurus (Wiegmann, 1834)

Identification: The Pacific Snake-eyed Skink is a small, up to 5.0 in., eugongylid lizard. This species is slender in shape with a long tail. Prominent gold paravertebral stripes extend through the tail. The sides are black with light stippling. Its venter is yellow. Osteoderms give a shiny smooth look to Hawai'i's only skink that lacks eyelids. Its legs and toes are long. Its center of distribution is the Pacific.

Introduction history and geographic range: The Pacific Snake-eyed Skink is presumed to have colonized the Hawaiian Islands as a stowaway with the

Photo courtesy of Gerald McCormack.

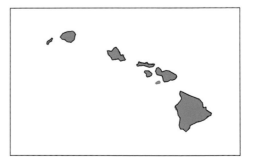

indigenous people of Hawai'i. In the United States, this species occurs on the Hawaiian islands of Hawai'i, Kaho'olawe, Kaua'i, Lāna'i, Maui, Moloka'i, Ni'ihau, and O'ahu.

Ecology: In Hawai'i, the Pacific Snake-eyed Skink is very common in rocky coastline habitat of small islands and can be abundant; however, it tends not to be found with other skink species. Individuals will climb about trees and rocks and forage on the ground as well. It is diurnal in activity and active throughout the year. The nesting season has not been identified in Hawai'i; however, clutch size of two eggs is typical, and females may share nesting sites. Eggs hatch in about four to seven weeks. Hatchlings measure less than 2.0 in. This skink is an insectivore. It is thought that beachfront development has negatively impacted this species.

COPPER-TAILED SKINK

Emoia cyanura (Lesson, 1830)

Identification: The Copper-tailed Skink is a small, up to 5.0 in., eugongylid lizard. Slender in build with a long tail, the Copper-tailed Skink is easily identified by a vertebral stripe and two dorsolateral white stripes. Background color is light brown. The tail is copper or light green in color. Its venter is white.

Photo courtesy of Gerald McCormack.

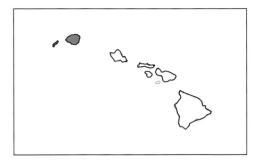

Osteoderms give a shiny smooth look to this species. Its center of distribution is the Pacific, where it is found on south and southwestern islands.

Introduction history and geographic range: This species is presumed to have colonized the Hawaiian Islands as a stowaway with the indigenous people of Hawai'i. In the United States, the Copper-tailed Skink occurs on the Hawaiian island of Kaua'i. The Copper-tailed Skink is known from one isolated population in the Po'ipu area of Kaua'i.

Ecology: Little is known about the ecology of the Copper-tailed Skink in Hawai'i. It is diurnal and, typically, two eggs are laid at a time and buried under leaf litter. This species is an insectivore. The Cattle Egret has been identified as a major predator of this skink in Hawai'i. Interestingly, both the Copper-tailed Skink and its close relative, the Azure-tailed Skink, are rare in Hawai'i. The reasons for their poor colonization success remain unknown; however, predation by the introduced mongoose and competition by the Plague Skink, both of which appeared in Hawai'i near the turn of the century, are implicated in the demise of the once common Copper-tailed Skink.

Azure-tailed Skink
Emoia impar (Werner, 1898)

Identification: The Azure-tailed Skink is a small, up to 5.0 in., eugongylid lizard. Slender in build with a long tail, the Azure-tailed Skink is easily identified by a vertebral stripe and two dorsolateral white stripes. Background color is light brown. The tail is blue in color. Its venter is gray to white. Osteoderms give a shiny smooth look to this species. There is also a bronze morph. Its center of distribution is the Pacific, where it is found on south and southwestern islands.

Introduction history and geographic range: It is presumed to have colonized the Hawaiian Islands as a stowaway with the indigenous people of Hawai'i. In the United States, the Azure-tailed Skink occurs on the Hawaiian islands of Hawai'i, Kaua'i, Maui, Moloka'i, and O'ahu. However, these records predate

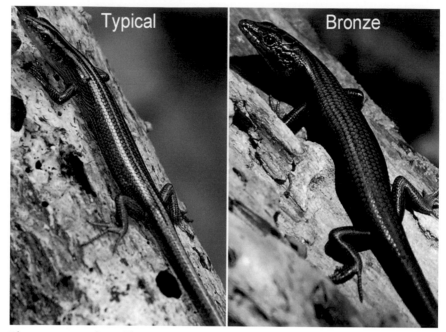

Photo courtesy of Gerald McCormack.

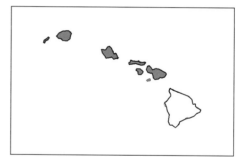

the 1950s, when this skink was commonly encountered and widespread. No contemporary records for the Azure-tailed Skink exist for Hawai'i.

Ecology: Little is known about the ecology of this diurnal insectivore in Hawai'i, where it was once common in mid-elevational habitats. Interestingly, both the Azure-tailed Skink and its close relative, Copper-tailed Skink, are rare in Hawai'i. The reasons for their poor colonization success remain unknown; however, predation by the introduced mongoose and competition by the Plague Skink, both of which appeared in Hawai'i near the turn of the century, are implicated in the demise of the once common Azure-tailed Skink. To that end, the Azure-tailed Skink is often abundant on South Pacific islands lacking a robust skink assemblage.

PLAGUE SKINK
Lampropholis delicata (De Viss, 1888)

Identification: The Plague Skink is a small, up to 5.0 in., eugongylid lizard. Slender in build with a long tail, the Plague Skink is brown in dorsal color. A longitudinal black band is present that may or may not be bordered on either side in white. Osteoderms give a shiny smooth look to this species whose body has a variable metallic luster. Its center of distribution is Australia.

Introduction history and geographic range: Arrival of this species in the United States was associated with human activities, where it was reported from O'ahu, Hawai'i, around 1900. In the United States, the Plague Skink occurs on the Hawaiian islands of Hawai'i, Kaua'i, Maui, Moloka'i, and O'ahu.

"*Lampropholis delicata* (De Vis, 1888), Dark-flecked Garden Sunskink" by Ian R McCann is licensed under CC BY 4.0. https://collections.museumsvictoria.com.au/species/12406.

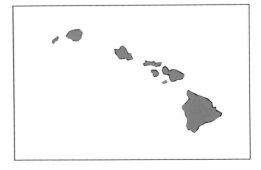

Ecology: The most common skink species in Hawai'i and a habitat generalist, the Plague Skink is found in many kinds of human-disturbed habitats and ecotones between secondary forest and open habitat, from sea level to mid-elevation. This species avoids heavy shade. Individuals are terrestrial and diurnal. The Plague Skink is a fecund species, laying up to seven eggs. Females may share a nesting site. Most abundant on O'ahu, adults are smaller and produce fewer eggs than on other islands of Hawai'i. This species is an insectivore. The Moth Skink is rare when in syntopy with the Plague Skink. Both the Copper-tailed Skink and Azure-tailed Skink, once abundant in Hawai'i, became scarce concomitant with the arrival of the Plague Skink. It remains to be seen what role, if any, the Plague Skink has played in the demise of these skink species.

Sun Skinks: Mabuyidae

BROWN MABUYA
Eutropis multifasciata (Kuhi, 1820)

Identification: The Brown Mabuya is a large, up to 10.0 in., mabuyid lizard. The body is drab olive to bronze in color. A yellow dorsolateral stripe of variable brightness is present. Its center of distribution is Asia.

Introduction history and geographic range: Arrival of this species in the United States was associated with the pet trade, where it was reported from

"Many-lined Sun Skink Mabuya multifasciata Singapore DSCF6795 (5).JPG" by Dr. Raju Kasambe is licensed under CC BY-SA 3.0. https://commons.wikimedia.org/wiki/File:Many-lined_Sun_Skink_Mabuya_multifasciata_Singapore_DSCF6795_(5).JPG.

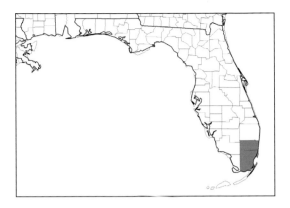

Coconut Grove, Miami-Dade County, in 1999 after being present since 1990. In the United States, the Brown Mabuya occurs in Miami-Dade County, Florida.

Ecology: The Brown Mabuya is abundant on the lushly planted grounds of the Kampong estate and in the surrounding neighborhood of Miami, Florida. Terrestrial and diurnal, individuals are associated with sunny open areas with mulch borders. Courtship may take place in March. This species is a livebearing species. The Brown Mabuya is an omnivore. On the grounds of the Kampong, the Brown Mabuya is syntopic with the native Florida Reef Gecko, *Sphaerodactylus notatus,* and likely encountered in the leaf litter. Its susceptibility to predation by the Brown Mabuya is unknown. Racers, present on the site, are a likely predator.

AFRICAN FIVE-LINED SKINK

Trachyleopis quinquetaeniata (Lichtenstein, 1823)

Identification: The African Five-lined Skink is a medium-sized, up to 8.0 in., mabuyid lizard. The body is brown in color. Five longitudinal stripes run along the body. The lateral stripes extend to the upper lips. The venter is gray to white. Adult males maintain only a light lateral stripe. Throat, upper lips, and sides of neck are white, and the tail is yellowish throughout. Juveniles have a bright blue tail and a strongly defined pattern, reminiscent of the native Common Five-lined Skink, *Plestiodon fasciatus.* Its center of distribution is Africa.

Introduction history and geographic range: Arrival of this species in the United States was associated with the pet trade. It was reported from Port St. Lucie, St. Lucie County, Florida, in 2010 where it had probably existed since 2005. In California, this species was first detected in Glendora, Los Angeles County, in 2016, collected in 2018, and reported as established in 2019. In the United States, the African Five-lined Skink occurs in St. Lucie County, Florida, and Los Angeles County, California.

"Rainbow Skink (Trachylepis quinquetaeniata) female . . ." by Bernard Dupont is licensed under CC BY-SA 2.0. https://www.flickr.com/photos/berniedup/50285288077.

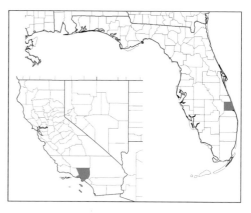

Ecology: The African Five-lined Skink is abundant in its limited area in Port St. Lucie, where it is associated with cement slabs, curbs, parking lots, and along buildings. This species is diurnal in its activity. Escape for most individuals is under ground cover. This species lays eggs and is an insectivore. Little is known about this species in Florida, although the isolation of this colony should make eradication feasible.

Sphenomorphid Skinks: Sphenomorphidae

MOTH SKINK

Lipinia noctua (Lesson, 1830)

Identification: The Moth Skink is a typically small, up to 4.0 in., sphenomorphid lizard. A distinctive bright yellow spot is present behind its head that connects to a faded mid-dorsal stripe. The rest of its dorsum and tail is light brown in color and the sides are dark. Osteoderms give a shiny smooth look to this species. Its center of distribution is the South Pacific.

Photo courtesy of Gerald McCormack.

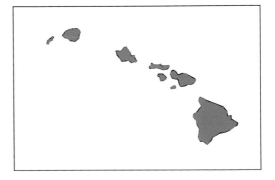

Introduction history and geographic range: The Moth Skink is presumed to have colonized the Hawaiian Islands as a stowaway with the indigenous people of Hawai'i. In the United States, this species occurs on the Hawaiian islands of Hawai'i, Kaua'i, Maui, Moloka'i, and O'ahu.

Ecology: The Moth Skink readily inhabits rock walls and the leaf litter of yards as well as around and on the root systems of large trees. It is principally diurnal but will on occasion forage after dark. An average of two young are produced through live birth. It is an insectivore. Predators include mice, rats, Mongoose, two species of bulbul, and the Cattle Egret. Once considered rare, the Moth Skink was more recently found to be more widespread; however, it is seldom abundant and is rare where it is found with the Plague Skink. The extent to which the Plague Skink plays a causative role in the rarity of the Moth Skink is unknown.

Ameivas, Whiptails, Racerunners, and Tegus: Teiidae

GIANT AMEIVA

Ameiva ameiva (Linnaeus, 1758)

Identification: The Giant Ameiva is a medium-sized, up to 18.0 in., teiid lizard. Among adults, males are larger than females. The skull is long and pointed. Adults may be green throughout or posteriorly. In the latter form, light brown with black stippling is present from the head variably past the shoulders. Dark-bordered white spots on the sides grade to blue dorsoventrally. Among adults, green is vibrant and generally more extensive in males than in females. The

"Amazon Racerunner—Giant Ameiva" by vil.sandi is licensed with CC BY-ND 2.0.
https://www.flickr.com/photos/38053299@N08/14669743785

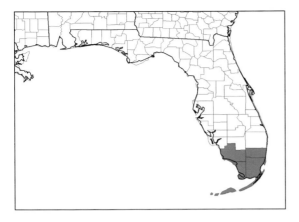

venter is blue in males and white in females. Jowls are well developed in adult males. Hatchlings are gray to brown and striped in green. Its center of distribution is Central and South America.

Introduction history and geographic range: Arrival of this species in the United States was associated with the pet trade, where it was reported from Miami, Miami-Dade County, Florida, in 1957 and more specifically in 1958. It has been established since at least 1954. In the United States, the Giant Ameiva occurs in extreme southern Florida.

Ecology: The Giant Ameiva is strongly associated with open habitat and nearby cover in the form of shrubs, weedy cover, and rubble. This species is terrestrial, strongly diurnal, and active throughout the year. Individuals are especially active during the midmorning and briefly in the late afternoon. Activity occurs between 8:00 am and 8:00 pm in July and 11:00 am and 4:00 pm during November–January. They will modify existing burrows or make their own burrows in which they sleep or retreat if frightened. The Giant Ameiva is very wary and very fast. When active, if not basking, the Giant Ameiva will be seen actively foraging, stopping here and there to scrape or dig with its forefeet and to poke the ground with its nose. Females and juveniles are less often seen foraging away from cover. The Giant Ameiva lays eggs; however, neither the clutch size nor the frequency has been reported in wild-caught individuals in Florida. Invertebrates are eaten as well as eggs of the Brown Anole. The native Eastern Six-lined Racerunner is ecologically very similar to the Giant Ameiva, placing the two species in potential competition and perhaps placing the racerunner at risk of predation by its larger relative. The Brown Anole seems likely at risk of predation at life stages beyond the egg. Racers eat young Ameivas. The Dusky Giant Ameiva hybridizes with the Giant Ameiva in one site in Pinecrest. This area represents a historical site for the Giant Ameiva. It is unknown whether the appearance of the Dusky Giant Ameiva was the result of an independent introduction or expansion.

Dusky Giant Ameiva
Ameiva praesignis (Baird and Girard, 1852)

Identification: The Dusky Giant Ameiva is a medium-sized, up to 24.0 in., te-iid lizard. Among adults, males are larger than females. The skull is long and pointed. Adult males are dark charcoal gray with transverse dorsal rows of yel-lowish-white spots that grade to blue laterally and on the rear legs. The venter is blue. Jowls are well developed in adult males. Adult females are comparatively lighter gray anteriorly than males and usually retain faded striping posteriorly. The venter is white. Hatchlings are gray to brown and striped in green. Its center of distribution is Central and South America.

"Dusky Giant Ameiva (Ameiva praesignis)" by Colin Meurk is licensed under CC BY-SA 4.0. https://www.inaturalist.org/photos/5518236.

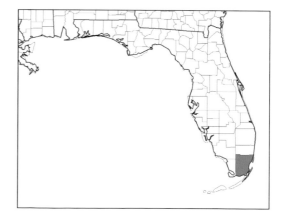

Introduction history and geographic range: Arrival of this species in the United States was associated with the pet trade, where it was reported from Hialeah and Miami, Miami-Dade County, Florida, in 1966. It was introduced in Hialeah in 1964 and in Miami before 1964. A separate introduction on Key Biscayne, Miami-Dade County, was derived from a zoological park and reported in 1983. In the United States, the Dusky Giant Ameiva occurs in extreme southern Florida.

Ecology: The Dusky Giant Ameiva is strongly associated with open habitat and nearby cover in the form of shrubs, weedy cover, and rubble. This species is terrestrial, strongly diurnal, and active throughout the year. Individuals are especially active during the midmorning and briefly in the late afternoon. Activity occurs between 8:00 am and 8:00 pm in July and 11:00 am and 4:00 pm during November–January. They will modify existing burrows or make their own burrows in which they sleep or retreat if frightened. The Dusky Giant Ameiva is very wary and very fast. When active, if not basking, the Dusky Giant Ameiva will be seen actively foraging, stopping here and there to scrape or dig with its forefeet and to poke the ground with its nose. Females and juveniles are less often seen foraging away from cover, and sexual pairs may forage together. The Dusky Giant Ameiva lays eggs; however, neither the clutch size nor the frequency has been reported for wild-caught individuals in Florida. Two hatchlings were caught on 7 May. Various invertebrates are eaten as well as eggs of the Brown Anole. Racers eat young ameivas. The native Eastern Six-lined Racerunner is ecologically very similar to the Dusky Giant Ameiva, placing the two species in potential competition and perhaps placing the racerunner at risk of predation by its much larger relative. The Brown Anole seems likely at risk of predation at all life stages. The Dusky Giant Ameiva hybridizes with the Giant Ameiva in one site in Pinecrest. This area represents a historical site for the Giant Ameiva. It is unknown whether the appearance of the Dusky Giant Ameiva was the result of an independent introduction or expansion.

GIANT WHIPTAIL
Aspidoscelis motaguae Sackett, 1941

Identification: The Giant Whiptail is a medium-sized, up to 12.0 in., teiid lizard. Among adults, males are larger in body length than females. The skull is long and pointed. Among adults, the overall body color is golden brown. Much of the dorsum, exclusive of head and neck, and rear legs are extensively covered in irregular yellow spots. The sides are darker, the tail grades to a light red, and the venter is blue. Females are paler than males. Its center of distribution is Central America.

Photo courtesy of Vicente Mata-Silva.

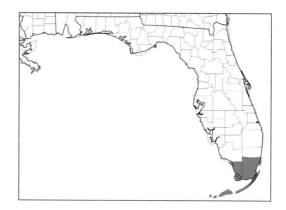

Introduction history and geographic range: Arrival of this species in the United States was associated with the pet trade, where it was reported from Miami and Opa-Locka, Miami-Dade County, Florida, in 1995. The Miami colony had been in existence for at least eight years prior to the report, and the Opa-Locka colony was at least 20 years old at the time of its report. In the United States, the Giant Whiptail occurs in two isolated sites in extreme southern mainland Florida.

Ecology: The Giant Whiptail is strongly associated with open habitat with nearby cover in the form of shrubs, weedy cover, and rubble. It can be found along canals and sidewalks bordered by shrubbery. This species is terrestrial, strongly diurnal, and active throughout the year. Individuals are especially active during the midmorning and seldom so on overcast days. They will modify existing burrows or make their own burrows in which they sleep or retreat if frightened. The Giant Whiptail is an active foraging species, moving about haltingly to scrape or dig for food with its forefeet and to poke the ground with its nose. Foraging generally takes place in close proximity to cover. Three sets of eggs were present in females captured in July. Eggs had presumably been laid prior to their capture. Hatchlings measure about 4.5 in. Beetles and ants make up much of its diet in Miami. The native Eastern Six-lined Racerunner is ecologically very similar to the Giant Whiptail, placing the two species in potential competition and perhaps placing the racerunner at risk of predation by its larger relative. The Brown Anole seems likely at risk of predation at life stages beyond the egg.

Rainbow Whiptail

Cnemidophorus lemniscatus (Linnaeus, 1758)

Identification: The Rainbow Whiptail is a medium-sized, up to 12.0 in., teiid lizard. Among adults, males are larger than females. Aptly named, this lizard shows many colors. Adult males have a broad grayish-brown dorsal stripe that extends from the head onto the tail. The edges of the stripe are darker brown

Photo by Suzanne Collins.

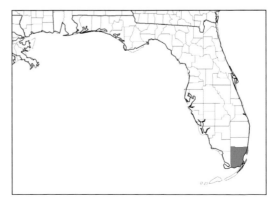

and in turn defined by a bright lime dorsolateral stripe. The sides of the body range from green to greenish yellow to bright yellow with yellow spots. The sides of the head, throat, and anterior aspect of the forelimbs are turquoise. The tail is a vibrant light green. The venter is gray. Females have faded yellow stripes on a dull green to brown background. A faded green ventrolateral stripe extends onto the tail. The rear legs are covered in light spots. The head is a faded yellow-orange, and the venter is white. Hatchlings are striped. Its center of distribution is Central and South America.

Introduction history and geographic range: Arrival of this species in the United States was associated with the pet trade, where it was reported from Hialeah, Miami-Dade County, Florida, in 1966. This colony did not survive. The species was subsequently reported from northern Miami-Dade County in 1983. These pet trade–derived colonies have waxed and waned since the 1970s but appear to be one continuous population. Several lines of evidence point to Colombia as the source or primary source of the Florida populations. In the United States, the Rainbow Whiptail is known to occur in the northern portion of Miami-Dade County in southern Florida.

Ecology: The habitats of the Rainbow Whiptail in Florida are sparsely vegetated vacant lots near warehouses and along railroad tracks. This species is terrestrial, strongly diurnal, and active throughout the year. Parthenogenetic populations exist in this species; however, the Florida population is composed of both sexes. At least two clutches averaging two eggs can be produced annually. Hatchlings measure about 4.5 in. The Rainbow Whiptail forages actively for invertebrates, especially ants and beetles, and vegetation in the form of leaves and flowers. The native Eastern Six-lined Racerunner is ecologically very similar to the Rainbow Whiptail, placing the two species in potential competition.

ARGENTINE GIANT TEGU

Salvator merianae Duméril and Bibron, 1839

Identification: The Argentine Giant Tegu is a large, up to 54.0 in., teiid lizard. Among adults, males are larger than females. Juveniles and adults are for the most part black and white of varying extent and intensity. Shades of light brown or gray may be present. Bands of white spots cross the dorsum. A white dorsolateral stripe or string of spots is typically well formed. A black stripe connecting the eye to the ear is often present, as is a thick lateral black band running from below and behind the ear to midway along the body. The venter is typically white and black. Exclusive to the Florida City, Florida, population, the venter of some individuals varies from faded to bright orange to red with black markings. The snouts of the Florida City tegus are longer and attenuated compared with the blockier and shorter snouts of the central Florida population. Jowls

"Brazil-01314—Many Lizards . . ." by Dennis Jarvis is licensed under CC BY-SA 2.0.
https://www.flickr.com/photos/archer10/48989978746/.

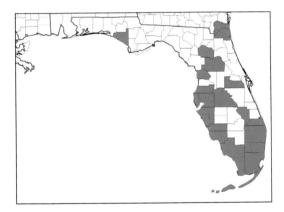

are well developed in adult males. Hatchlings are bright green anteriorly with transverse black bands across the dorsum. A line or string of white dorsal spots runs dorsolaterally along the body, and the tail is banded in black and white. Its center of distribution is South America.

Introduction history and geographic range: Arrival of this species in the United States was associated with the pet trade, where it was reported initially from Hillsborough and Polk counties, Florida, in 2007. These lizards were derived from Paraguayan stock introduced in the early 2000s. A second colony, also derived from released animals from a single source, was reported from Florida

City, Miami-Dade County, in extreme southern Florida, in 2007. Individuals were released at that site in the mid-1990s.

Ecology: In the United States, the Argentine Giant Tegu occurs in two as yet unconnected colonies in peninsular Florida. Disturbed secondary growth habitat, successional shrubland, and agricultural areas are quickly colonized; however, it is capable of colonizing xeric habitats as well, readily using Gopher Tortoise burrows. This species is terrestrial, strongly diurnal, and seasonal in its activity. Individuals sleep in naturally occurring burrows and burrows of their own making, and they are not averse to sharing their burrows with one another. Individuals of the South Florida population progressively spend less time overwintering in their burrows. Individuals living along the periphery wander farther than those in the core of the population; however, activity ranges are similar between the two areas. Mating commences in the spring. Primarily in early summer, a single clutch of 20–40 eggs is laid in a nest that is constructed and guarded by the female. Most young hatch two months later, in August and September. Hatchlings typically measure 7.0–10.0 in.

Like the other teiid lizards of Florida, the Argentine Giant Tegu is an active forager. It is an omnivore whose diet includes fruits, invertebrates, and vertebrates. Hatchlings are subject to a wide range of predators, some of which may be negatively impacted by adults and potentially by the Gold Tegu. Adults probably face few predators in Florida, although along South Florida canals individuals might be subject to the depredations of the Burmese Python. Colonization by the Argentine Giant Tegu places other resident species at risk as well. The use of Gopher Tortoise burrows was shown to inhibit use by Gopher Tortoises when occupied by multiple individuals of Gray's American Spiny-tailed Iguana, and the Argentine Giant Tegu could present a similar risk. Other burrowing species, such as the Florida Burrowing Owl, as well as some of the commensal species associated with Gopher Tortoise burrows, could be negatively impacted by the presence or depredations of this lizard. For instance, an approximately 47.0 in. tegu was observed shaking an approximately 52.0 in. Red Cornsnake as the lizard entered a Gopher Tortoise burrow. Ground-nesting birds could also be at risk of disturbance or predation on nests. Juvenile Gold Tegus could be at risk of predation by these large and fast lizards.

From a financial standpoint, attraction to sweet fruits by the Argentine Giant Tegu places some low-growing crops, such as strawberries, at risk. In southern Florida, where population sizes are high, and throughout the agricultural belt, a financial impact presents a genuine concern. Related to the loss of product through foraging is a health and human safety concern of potential for transmission of *Salmonella*, a combined concern not to be taken lightly. Perhaps not surprisingly, a report generated by a panel of scientists and resource managers identified the Argentine Giant Tegu as likely the most threatening of Florida's exotic reptiles.

Dispersal of the Argentine Giant Tegu from southern Florida follows a northwestern direction. Connection with the central Florida population is inevitable and will place at substantial risk many members of the sandy upland fauna. Encouragingly, recent research has developed and tested practical, easy-to-apply tracking plot methods that are applicable during approximately six months of the year. These survey methods offer inexpensive means to track range expansion and assess population sizes and should be used to detect and monitor population trends of this tegu species in the central Florida area.

GOLD TEGU

Tupinambis teguixin (Linnaeus, 1758)

Identification: The Gold Tegu is a medium-sized, up to 30.0 in., teiid lizard. Individuals are sleek and slender. The head is long and narrow and not deep. The tail is more than one-half the tegu's total length. The gold ground color is stippled in black, and broken black bands pass from the sides over the dorsum. The tail is banded in black and gold. Among adults, males are darker in color. Its center of distribution is South America.

Photo by Suzanne Collins.

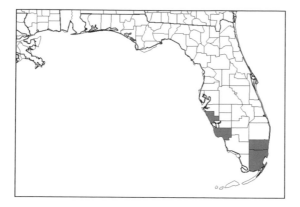

Introduction history and geographic range: Arrival of this species in the United States was associated with the pet trade. The Gold Tegu was first reported from Key Biscayne, Miami-Dade County, Florida, in 2010, from a record taken in 1995. However, the first established colony of the Gold Tegu was published in 2017 from records made during 2008–2016 in Florida City, Miami-Dade County. In the United States, the Gold Tegu occurs in one area of southwestern Miami-Dade County, Florida.

Ecology: In Florida, the Gold Tegu inhabits heavily disturbed, well-vegetated habitat. This species is terrestrial and strongly diurnal. In South America, the eggs hatch as rains begin in June or July, approximately 150–180 days after having been laid. The nest is not guarded by the female. Like the other teiid lizards of Florida, the Gold Tegu is an active forager. It is an omnivore whose diet includes fruits, invertebrates, and vertebrates. Hatchlings are subject to a wide range of predators, some of which may be negatively impacted by adults and by the Argentine Giant Tegu. Ground-nesting birds could be at risk of disturbance of or predation on nests by the Gold Tegu.

Monitor Lizards: Varanidae

NILE MONITOR

Varanus niloticus Linnaeus *in* Hasselquist, 1766

Identification: The Nile Monitor is a large, up to 94.5 in., varanid lizard. Adult color pattern can vary in intensity. In general, the background color is a dull gray-green. The head is long and slender and may have a black bar running from behind the eye to just above the ear. Rows of yellow spots run crosswise over the sides and dorsum. The legs are lightly spotted. The venter is dull yellow with dark markings. The tail is long, laterally compressed, and overall dark with light crossbands. More strikingly patterned than adults, hatchlings are dark gray

Photo courtesy of Bill Love.

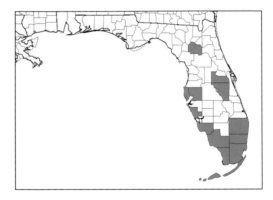

above and heavily flecked in white. Rows of yellow spots run crosswise over the sides and dorsum. These rows of spots are more or less connected to one another on the tail. The throat is light yellow and flecked in black. Its center of distribution is Africa.

Introduction history and geographic range: Arrival of this species in the United States was associated with the pet trade, where it was first reported from Cape Coral, Lee County, Florida, in 2004. This colony had been in existence since at least the early 1990s. In the United States, the Nile Monitor is well established in southern Florida, especially in areas of Lee, Miami-Dade, and Palm Beach counties.

Ecology: In southern Florida, the Nile Monitor is associated primarily but not exclusively with disturbed habitat near water. Habitats include mangrove forest, canals, and borrow pits. This species might best be considered semiaquatic-terrestrial when adult, and extensive climbing is associated with hatchlings and juveniles. This species is strongly diurnal and active throughout the year in southern Florida. Individuals sleep in a burrow of their own making or in a modified existing burrow where they also retreat when frightened. An alarmed Nile Monitor can run quickly and escape into water, a burrow, or up a tree. If cornered, an adult Nile Monitor can be dangerous.

Mating has been observed in June in Palm Beach County, a gravid female in August in Cape Coral, and hatchlings in November, January, and February in Palm Beach County. Clutch size is unknown for Florida populations, but clutch sizes of up to 60 eggs can be produced by captive females. Incubation may take up to 10 months. Hatchlings measure approximately 11.0 in. Sexual maturity can be reached in two or three years.

This species is an active forager and is carnivorous. Juveniles have sharp teeth to catch their prey. Posterior teeth of adults are blunt in shape and used for breaking or crushing their prey. Sharp claws are used for digging and for rendering prey. Olfaction is well developed in this species, and its long-forked tongue assists in its search for food, such as a carcass or a hidden nest of eggs. Adults and eggs of the Brown Anole are reported from its diet in Florida. Hatchlings are likely subject to a wide range of vertebrate predators in Florida. Subadults and adults likely face few predators, perhaps the American Alligator and the Burmese Python. The Nile Monitor places many of Florida's sensitive species at potentially great risk. Eggs of sea turtles, Diamond-backed Terrapin, *Malaclemys terrapin,* Gopher Tortoise (and hatchlings), American Alligator, and American Crocodile, *Crocodylus acutus,* as well as eggs of the exotic Spectacled Caiman are all at risk of predation. Considered an airstrike hazard to military aircraft, monitors were actively removed from Homestead Air Reserve Base. Dispersal of the Nile Monitor from southern Florida can be expected along both coasts with mangrove, riverine, and canal systems to its liking.

PART 5

SNAKES

Burmese Python Colonization
and Eradication

A Stitch in Time Unnecessarily Missed in the Everglades

MICHAEL R. ROCHFORD

Why have eradication efforts of the Burmese Python, *Python bivittatus,* failed in the Everglades? I believe that it helps to reflect on the Burmese Python situation when the issue first captured attention, because the odds of success in controlling the species were greatest in the beginning. Burmese Pythons firmly established themselves in southern Florida by the early 2000s. Interagency efforts were ramping up to control or eradicate the new poster child for invasive species. More than 150 pythons were removed in 2006 alone. The number removed each year grew exponentially during that decade. Fast forward to late 2020 and two agencies boast of having removed 5,000 Burmese Pythons in approximately three years, and the species persists in the Everglades.

The status of the Northern African Python, *Python sebae,* in the Bird Drive Basin of Miami-Dade County serves as an analogue to the management history of the Burmese Python. Fewer than 50 Northern African Pythons have been reported since 2001, and yet despite sustained survey and removal pressure the population persists. This area is less than 1 percent of the size of Everglades National Park. So, when did we miss our chance to extirpate the Burmese Python from the Everglades? I argue that we missed the boat by 2006 while biologists had struggled to convince stakeholders that the python was even a problem.

Strong incentive existed to dismiss the gravity of the situation. Python breeders feared their livelihoods would be impacted, and breeders held influence over many hobbyists. For some agencies, official recognition of the existence of a wild python population would have meant having to allocate funds to combat the emerging issue, and to the average person this may have sounded like a far-fetched idea.

Denial of the problem was the easiest position to take. The burden of proof was on biologists, and until they produced sufficient evidence, it seemed as though any argument contrary to the existence of Burmese Pythons in the Everglades gained traction. Naysayers argued that pythons could never make it in the Everglades because Red Imported Fire Ants, *Solenopsis invicta*, would eat their eggs, and that Florida is too cold for pythons to survive, let alone reproduce. Finally, if pythons were there in numbers estimated by biologists, wouldn't people find more?

Why weren't people finding more? Shouldn't any snake longer than 10 feet stick out like a sore thumb? Yes, if you find one crossing a road or levee. However, after radio-tracking them and seeing how cryptically they live most of their lives, you quickly understand that they are the proverbial needle in the 1.5-million-acre haystack. Put simply, any person attempting to convey authority about the invasive Burmese Python would have greater credibility by first having radio-tracked them, because this sort of research is used to more accurately estimate the difference between how many pythons are seen and how many actually exist in the wild.

The difficulty of this kind of work is not to be taken lightly. The River of Grass is not a manicured lawn, it is sawgrass and native cattails that reach over your head. Just add water, deep mud, other aquatic vegetation, mosquitoes, alligators, heat, and humidity. If you can hike 1 mi./hr. in this marsh, you are doing well. Air temperatures and humidity, both in the 90s, make conditions dangerous for sustained physical activity in the summer, but if you want to find pythons on their turf this has to be endured. Some areas are accessible by hiking from a road, and others can be reached by airboat or motorboat, but access to most sites requires helicopter assistance.

What can happen in the marsh itself? I spent 10 years performing field research on pythons in the Everglades, and once while tracking a 12-foot python with two coworkers, we experienced some trouble. Our telemetry equipment can pinpoint the snake's exact location, which we often walk past at first, as we did that day. We were in periphyton-covered water up to our calves and could not see even half an inch below the surface. The third person in line let out a blood-curdling scream. We turned to see the python's jaws attached to his bare leg as he reached down and grabbed it by the neck. In the process of releasing himself from the snake, he tore small lacerations where each of more than 100 teeth had penetrated his flesh. There we were, three experienced python trackers searching with the help of equipment for a 12-foot python we knew was nearby, yet all of us had missed it and one of us was a bloody mess.

Similar situations, sans bite, played out hundreds of times during our research. Pythons spend a lot of time submerged in the water. When on dry ground, they usually hide under leaf litter or other vegetation. An even loosely coiled 10-foot Burmese Python can take up surprisingly little room, and individuals often seek refuge in subterranean pockets carved in limestone under tangled root systems. Only during very cold conditions do they bask and make themselves visible. These data are difficult to come by and are necessary for sound assessments of populations.

Notwithstanding the difficulties in finding Burmese Pythons in natural habitat, right along the road they can still be tough to spot. For example, one night on the mowed shoulder of a road in Everglades National Park, I stopped with a friend while looking for snakes. We examined a culvert running under the road for signs of life when she said, "I think I'm standing on something." From experience, I immediately suspected a python. We carefully peeled back a thin layer of mowed grass to reveal a 10-foot python coiled quietly in the dark, demonstrating the python's ability to hide effectively in three inches of grass essentially in the open.

Most people would understand this if they had radio-tracked a few pythons in the River of Grass and most are understandably not afforded the opportunity. We should trust those who do get the chance. Biologists knew early that Burmese Pythons roamed the Everglades in large numbers and knew just how difficult detection of snakes could be and, therefore, how easy it would be to underestimate their population sizes. However, to sway policy is difficult in a world where one person's opinion is given equal weight to the expertise of others.

Consequently, by the time biologists succeeded in convincing stakeholders of the looming python threat, the window for eradication had passed. Pseudo-expert arguments delayed efforts and ensured pythons a place in the Everglades for many years to come, and at the expense of countless mammals, birds, and alligators that are consumed by this species. Serious funding for control efforts arrived too late. This outcome, of course, did not have to happen. I suggest that when new species are inevitably introduced, the extent of invasion should be questioned immediately. The initial stages of invasion are not the time for disbelief. Facts must win the day. If only one individual appears, initial rapid response efforts could successfully address the problem. Detection of more individuals deserves immediate attention and action. For every individual seen, how many are missed? Our experience with Burmese Pythons places that estimate at roughly one individual seen in surveys per 100 in existence. We refrain from overconfidence that makes it easy to believe you got them all. As a stitch in time, scraping together funds to eradicate a small population is several orders of magnitude easier and less expensive than allowing that population to grow too large for eradication or containment.

We failed the python test. Let's do better.

Snakes (Squamata)

Boas: Boidae

BOA CONSTRICTOR
Boa constrictor Linnaeus, 1758

Identification: The Boa Constrictor is a large, up to 140.0 in., boid snake. Among adults, males are smaller than females. The largest individual in a Florida study measured 98.8 in. This is a heavy-bodied snake. Regionally distinct forms exist, although the Florida population most resembles those of Colombia. Background color is brown and light to dark tan that may be variably flecked in black. The head is tan with a dark stripe running from just behind the eye to the end of its jaw and a dark head stripe running from the snout to the neck.

Photo by Suzanne Collins.

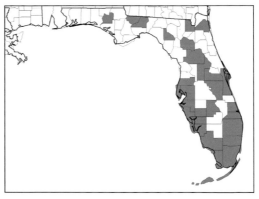

Thereafter, light to dark brown saddles of variable length continue through the tail, which grades to variable shades of orange-red. The venter is whitish and variably marked in black. Hatchlings are more strikingly patterned than adults. Its center of distribution is South America.

Introduction history and geographic range: Arrival of this species to the United States was associated with the pet trade, where it was first reported from Miami, Miami-Dade County, Florida, in 1994 and thought to have been established. This species was convincingly shown to be established in 2007 at the same site and thought to have occurred there since the 1970s. In the United States, the Boa Constrictor is well established at the Deering Estate and vicinity in Miami-Dade County, Florida.

Ecology: In southern Florida, the Boa Constrictor inhabits rockland pine, tropical hardwood hammock, and adjacent habitats. Individuals are active day and night and throughout the year in southern Florida. In southern Florida, live birth occurred in May and June, with litter sizes of 24 and 34 having been reported. In August and September, 86% of captures were young-of-the-year. Hatchlings are approximately 20.0 in. The Boa Constrictor is carnivorous and a powerful constrictor. In southern Florida, an individual captured in August had eaten a Virginia Opossum, *Didelphis virginiana*. Young are likely subject to a wide range of vertebrate predators in Florida, including the Eastern Indigo Snake, *Drymarchon couperi*, at the Deering Estate. Adults may be relatively predator-free in southern Florida. In light of its broad diet and potentially large body size, the Boa Constrictor places a wide range of vertebrates at risk through its depredations. Interestingly enough, however, unlike the Burmese Python, the Boa Constrictor has remained confined to a small, if thriving, colony.

Harmless Live-bearing Snakes: Natricidae

SOUTHERN WATERSNAKE
Nerodia fasciata (Linnaeus, 1766)

Identification: The Southern Watersnake is a medium-sized, up to 42.0 in., natricid snake. Among adults, males are smaller than females. Although variable, its dorsal background color is usually reddish brown with black crossbands. A distinct black stripe extending from the eye to the jaw is present. The venter may have blotches or vermiculations. The Southern Watersnake is a polytypic species comprising three subspecies: the Broad-banded Watersnake, *N. f. confluens*, Banded Watersnake, *N. f. fasciata*, and the Florida Watersnake, *N. f. pictiventris*. Collectively, they are found throughout much of the southeastern coastal United States, Florida, and Louisiana, and eastern portions of Arkansas and Texas.

Banded Watersnake, *N. f. fasciata*. "Photo 96774669" by Steve Taylor is licensed under CC BY 4.0. https://www.inaturalist.org/photos/96774669.

Florida Watersnake, *N. f. pictiventris*.
Photo by Janson Jones.

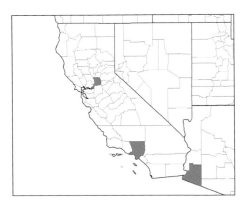

Introduction history and geographic range: In the United States, the Banded Watersnake extralimitally occurs near Lake Natoma, Folsom, Sacramento County, California, where it has been known since 1992. The Southern Watersnake, its subspecies not specified, is exotic to Imperial Dam, Yuma County, Arizona, where it was caught in 2015 and reported in 2018. The Florida Watersnake is exotic to Lake Machado, Harbor City, Los Angeles County, California, where it was reported in 2006.

Ecology: This species is found in shallow wetlands, well-vegetated canals, ponds, and lake edges. This snake can be abundant both in the southern Everglades and in Lake Machado. In southern Florida, litter size averages 16 young, and most are born in midsummer. Growth to maturity in southern Florida occurs in approximately one year after birth. In Florida, this species feeds primarily upon amphibians, especially frogs and toads, and will include fish in its diet. In Lake Machado, its diet was found to change from the exotic Mosquitofish when young, to exotic North American Bullfrogs when adult. Herons and egrets are predators of this species. The success of this snake in Folsom raises the concern that other similar habitats could be easily colonized, especially those with North American Bullfrogs. The Southern Watersnake fills a poorly exploited niche and places native species, including several listed species, at risk, such as the California Tiger Salamander, Foothills Yellow-legged Frog, *Rana boylii*, and Red-legged Frog.

COMMON WATERSNAKE
Nerodia sipedon (Linnaeus, 1758)

Identification: The Common Watersnake is a medium-sized, up to 42.0 in., natricid snake. Among adults, males are smaller than females. Although variable, its dorsal background color is usually dark brown with reddish bands and blotches that alternate near the posterior end of the snake. Dorsal markings may fade in old adults. The venter is white with extensive and pale to vivid red half-moons or may be a pale orange throughout. Juveniles contrast in pale gray background with black dorsal markings. The Common Watersnake is a polytypic species comprising four subspecies: the Northern Watersnake, *N. s. sipedon*, Lake Erie Watersnake, *N. s. insularum*, Midland Watersnake, *N. s. pleuralis*, and Carolina Watersnake, *N. s. williamengelsi*. Collectively, they are found throughout much of the eastern United States.

Introduction history and geographic range: In the United States, the Common Watersnake is exotic to Kaseberg Creek, Roseville, Placer County, California, where it was collected and reported in 2007. In Colorado, it has been reported from the Arikaree, Arkansas, and South Platte river drainages, the extent unclear.

Photo by Suzanne Collins.

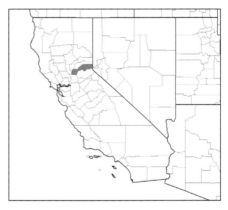

Ecology: This species is found in a wide range of lentic to slow-moving freshwater systems. Population size was estimated to be 348 individuals, and captures of males and females were similar in numbers. Of 119 snakes, six individuals were missing a portion of their tail. In Roseville, mating was observed in April, with live birth in August by one female. Smallest reproductive individuals were smaller in males (13.7 in. snout-vent length or SVL) than females (23.4 in. SVL). Individuals feed on amphibians, especially frogs and toads, and fish. Prey items recovered from the Roseville population were exotic North American Bullfrogs and native Pacific Chorus Frogs, *Pseudacris regilla.* Herons and egrets are predators of this species. The success of this snake in Kaseberg Creek raises the concern that other similar habitats, especially those with North American Bull-frogs, can be colonized. Furthermore, the Common Watersnake fills a poorly exploited niche and places native species, including several listed species, at risk, such as the California Tiger Salamander, Foothills Yellow-legged Frog and Red-legged Frog.

SHORT-HEADED GARTERSNAKE
Thamnophis brachystoma (Cope, 1892)

Identification: The Short-headed Gartersnake is a small, up to 18.0 in., natri-cid snake. Among adults, males are smaller, averaging 11.0 in., than females, which average 13.0 in. The head seems barely set apart from the body and ap-pears undersized. Background color is dark brown. A bright yellowish to tan vertebral stripe is usually present; however, in some individuals the vertebral stripe is either greenish brown or indistinct. A dull lateral stripe is present. The venter is tan. The natural geographic range of the Short-headed Gartersnake is restricted to unglaciated areas of southern New York and portions of western Pennsylvania.

Introduction history and geographic range: In the United States, the Short-headed Gartersnake is exotic to urban areas of Allegheny, Butler, and Erie coun-ties of Pennsylvania.

Ecology: The Short-headed Gartersnake prefers open habitats, such as old fields, grasslands, and meadows. Despite a very small geographic range, this species

Photo by Suzanne Collins.

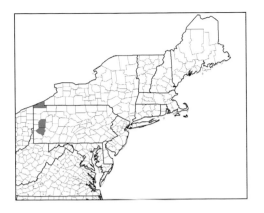

can be very abundant where readily found under flat cover. In Pennsylvania, individuals are active during April–October, with a peak in mid-June. A smaller August peak is associated with young-of-the-year. Mating takes place in the spring with males using sperm stored from the previous fall. This species gives live birth, and young are born in August. Litter size averages 9.2 young in Pennsylvania. Neonates are approximately 6.0 in. Males reach sexual maturity in one year, females in two years. Its diet primarily, if not exclusively, comprises earthworms. The status of its extralimital colonies is not well known.

Pythons: Pythonidae

BURMESE PYTHON
Python bivittatus Kuhi, 1820

Identification: The Burmese Python is a large, up to 226.0 in., pythonid snake. Among adults, males are smaller than females. One large female from southern Florida measured 191.7 in., and another large individual measured 179.9 in. This is a very heavy-bodied snake. Individuals are patterned in brown, tan or yellowish-tan, and white. The venter is white with some black mottling. The head has a distinctive dark brown arrowhead or spearhead-shaped mark with a light tan stripe in the center. A broad tan border runs from the nose along the upper jaw and passes through the top half of the eye. Its center of distribution is Southeast Asia.

Introduction history and geographic range: Arrival of this species in the United States was associated with the pet trade, where it was first reported established in Everglades National Park, Miami-Dade and Monroe counties, Florida, in 2000. Well established in the saline glades and mangroves at the southern end of the park, individuals had been reported there since the 1980s. The body size-distributions of the pythons and the profound population explosion reported

Photo courtesy of
Michael R. Rochford.

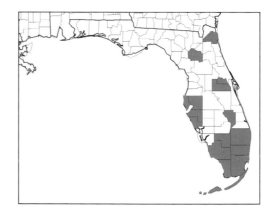

there and farther north in the park in the early 2000s and later onto the upper Florida Keys indicate that the saline glades colony was expansive in area and likely established by the early 1990s. Published models conflict on ultimate range of expanding populations; however, in the short term, stable populations are likely on the northern shore of Lake Okeechobee.

Ecology: In the United States, the Burmese Python occurs in wetland and upland habitats of the Everglades as well as in disturbed habitat in southern Florida. This species is active day and night and throughout the year in southern Florida. Movements are primarily diurnal between October and April, and primarily nocturnal between June and August. In the winter, males are more apt to be seen moving in the daytime in search of females. Males will also wander great distances. One radio-tracked Everglades male moved 43 miles. The Florida population is not behaviorally or physiologically adapted to withstand severely cold temperatures, as observed in the field and in experimental outdoor pens during a severe 2009–2010 winter cold spell in Florida.

In the Everglades, mating takes place in the winter, often with one female being courted by multiple males. Gravid females were collected between January and April. Clutch sizes ranged 21–85 eggs. Nests are constructed and guarded by the female. Two nests were found in May. In one nest, the hatching success was 92%. Hatchlings were seen in June. The Burmese Python hatches at a large body size of approximately 22.0 in. and grows quickly to a body and gape size that can handle very large prey. Consequently, small (e.g., Cotton Mouse, *Peromyscus gossypinus,* Northern House Wren, *Troglodytes aeodon,* juvenile American Alligator), medium (e.g., Bobcat, *Lynx rufus,* Pied-billed Grebe, *Podilymbus podiceps,* juvenile American Alligator), and large (White-tailed Deer, *Odocoileus virginianus,* Wood Stork, *Mycteria americana,* adult American Alligator) prey encompassing many taxa are eaten in southern Florida. One study documented severe declines in several mammal species in Everglades National Park following a population explosion of the Burmese Python. Along canals, the Burmese Python is a potential predator of the Nile Monitor, Argentine Giant Tegu, Green Iguana, and Spectacled Caiman. An array of avian and mammalian species are potential predators of small pythons. The American Alligator, a documented prey item of the Burmese Python, is also a documented predator of it.

Negative impacts associated with the Burmese Python in Florida affect native and exotic species. Its predator pressure has resulted in negative impacts on the food web in Everglades National Park. It also places several sensitive species at risk. Inclusion of deer in its diet places the Burmese Python in potential competition with the Florida Panther, *Puma concolor couguar.* In the Everglades, the Burmese Python negatively impacts the American Alligator through predation. It also has access to more prey species and is potentially subject to fewer predators than the American Alligator. A large Burmese Python is also a potential danger to humans, both directly through physical contact and indirectly through vehicle collisions.

Northern African Python
Python sebae (Gmelin, 1788)

Identification: The Northern African Python is a large, up to 192.9 in., pythonid snake. Among adults, males are smaller than females. In a Florida study a large male measured 110.6 in., and the largest female measured 192.9 in. This is a heavy-bodied snake. Individuals are variably blotched in tan, brown, and white. The head has a distinctive dark brown arrowhead or spearhead-shaped mark with a light tan stripe in the center. A distinctive triangular-shaped mark is present under each eye. A broad tan border runs from the nose along the upper jaw and passes through the top half of the eye. Thereafter, irregular-shaped dark brown blotches of variable size on a tan background continue through the tail. The venter is heavily mottled in black and white. Hatchlings are more strikingly patterned than adults. Its center of distribution is Africa.

Photo courtesy of Michael R. Rochford.

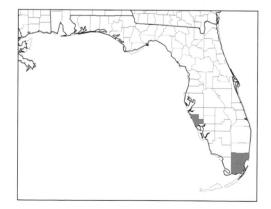

Introduction history and geographic range: Arrival of this species in the United States was associated with the pet trade, where it was first reported from a site known as the Bird Drive Basin in Miami, Miami-Dade County, Florida, in 2010, where individuals had been seen since 2002.

Ecology: In the United States, the Northern African Python occurs in a human-disturbed area adjacent to the eastern edge of Everglades National Park. Habitat of the Bird Drive Basin comprises residential, disturbed habitats, wetlands, and canals. Ideal habitat for the Northern African Python is permanent water with an upland connection. This species is presumably active day and night and is active throughout the year in southern Florida. In southern Florida, this species mates during December–April, and females are gravid during January–May. Females ranging 105.3–163.4 in. in size produced an average of 32.1 eggs, ranging from 11 to 47 eggs. Females guard a nest of their construction. A neonate measuring 26.4 in. was captured in the Miami colony in August. The Northern African Python is carnivorous and a powerful constrictor. In southern Florida, an approximately 28.0 in. juvenile was found to have eaten a Boat-tailed Grackle, *Quiscalus major*. Hatchlings are likely subject to a wide range of vertebrate predators in Florida, whereas adults may be relatively predator-free in southern Florida. In light of its diet and potentially large body size, the Northern African Python places a wide range of vertebrates at risk through its depredations. This colony is known from human-disturbed habitat, and considerable effort has been directed to its eradication since 2009; however, the extent of its range is not fully known.

Blindsnakes: Typhlopidae

BRAHMINY BLINDSNAKE
Indotyphlops braminus (Daudin, 1803)

Identification: The Brahminy Blindsnake is a small, up to 8.0 in., typhlopid snake. Easily mistaken for a worm, this thin and nondescript snake is uniform dark gray–dark brown-black. The head and neck are similarly thick, and the tail is short and pointed. Its center of distribution is Southeast Asia and through human mediation it has a pantropical distribution.

Introduction history and geographic range: Arrival of this species in the United States and subsequent dispersal events were associated with the ornamental plant trade. Reported first in Hawai'i, it was first found on O'ahu in 1930 and is now present on the Hawaiian Islands of Hawai'i, Kaho'olawe, Kaua'i, Lāna'i, Maui, Moloka'i, and O'ahu. In Arizona, it was first reported in 2013 from Phoenix, Maricopa County, from specimens collected in May and December of 2010. Other specimens had been collected from the site twice in 2007. In California,

Photo by Suzanne Collins.

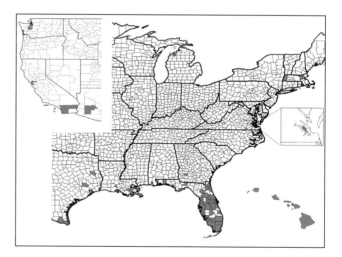

it was first reported in 2010 from Chula Vista, San Diego County, from specimens collected in 2006 and 2009. In Florida, it was first reported in 1983 from three separate sites in Miami-Dade County, where it had presumably been established since the 1970s. In Georgia, it was first reported in 2007 from Albany, Dougherty County, from specimens collected in 2007 and seen since 2005. In Louisiana, it was first reported from the Mid-City area of New Orleans, Orleans Parish, in 1994, where it had been found in 1993. In Massachusetts, it was first reported from a building in Boston, Suffolk County, in 1991 and again in 1995. In Texas, it was first reported in Brownsville, Cameron County, in 2000, and is established in an urban setting. Subsequently, it was reported in Nacogdoches County in 2007 and Hidalgo County in 2009. In Virginia, it was first reported from buildings in Newport News in 2002 although captured in 2000.

Ecology: In the United States, the Brahminy Blindsnake occurs in relatively mesic systems with most soils suitable for burrowing. In Florida, it appears to be restricted to mesic disturbed systems. In Hawaiʻi, individuals are found in gardens and moist valleys. In Louisiana, Massachusetts, Texas, and Virginia, colonies are associated with human habitation and found outdoors. Interestingly, one individual was collected from a palm boot nearly 6 feet above the ground. Under the boot, moist detritus was teeming with small invertebrates. The Brahminy Blindsnake is an all-female species that can produce between two and four elongate eggs at a time. The Brahminy Blindsnake is highly fossorial and drawn to termite nests for prey and to ant nests, such as those of the Florida Carpenter Ant, *Camponotus floridanus,* to prey on pupae. Such nests also provide protection from the elements and potential predators. In southern Florida, a Crested Anole defecated a Brahminy Blindsnake. Very little is known about the biology of this species in the United States, where its distribution is masked in part because of its fossorial nature and superficial resemblance to an earthworm.

PART 6

CROCODILIANS

Crocodilians (Crocodylia)

Alligators and Caimans: Alligatoridae

SPECTACLED CAIMAN

Caiman crocodilus (Linnaeus, 1758)

Identification: The Spectacled Caiman is a medium-sized, up to 94.0 in., alligatorid species. Adults taken from southern Florida typically measure less than 72.0 in. Among adults, males are larger than females. Adults are an overall green-gray in color and can darken in color in cold weather. Juveniles are yellow and speckled in black. The prominent bony ridge between the eyes is reminiscent of spectacles or eyeglasses. Although very wary by nature, it can nonetheless be a dangerous animal under the wrong circumstances. Its center of distribution is South America.

Photo by Suzanne Collins.

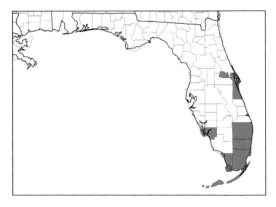

Introduction history and geographic range: Populations of this species in the United States were derived from the pet trade. The Spectacled Caiman was first noted in the United States in South Florida in 1966 but not known to be established. A possibly established colony in Miami, Miami-Dade County, Florida, was known in the 1950s, and reported in 1983. A Miami population dating back as early as 1960 was confirmed as established in 1980. Another Miami population, reported in 1983, was discovered in 1968. A third population, in Homestead, Miami-Dade County, first discovered in 1974, was reported in 1980. Subsequent dispersal through intentional release is common and often in disparate locations. In the United States, the Spectacled Caiman is found in portions of eastern central and southern mainland Florida. In response to legal protection of the American Alligator, interest soared in the Spectacled Caiman for hides and pets.

Ecology: The Homestead Air Reserve Base and surrounding canals and rock pits have long been secure areas for this species. To date, it has not colonized the Everglades. The female constructs a nest in the form of a mound consisting of vegetation and earth. The nest of up to 40 eggs is guarded aggressively by the female. Incubation lasts approximately 90 days, and hatchlings measure about 8.0 in. The young remain with the parents for more than one year. A nest of the Peninsula Cooter, *Pseudemys peninsularis,* was found in the nest of a Spectacled Caiman. The Spectacled Caiman is a carnivore. In southern Florida, young eat small fish, frogs, tadpoles, and insects. Adults eat mostly fish. Both native and exotic fishes are subject to its predation. Its interactions with the American Alligator and the Burmese Python in southern Florida warrant study.

References

Bartlett, R. D., and P. P. Bartlett. 1999. *A Field Guide to Florida Reptiles and Amphibians.* Gulf Publishing, Houston, TX. 280 pp.

Bartlett, R. D., and P. P. Bartlett. 2011. *Florida's Frogs, Toads, and Other Amphibians: A Guide to Their Identification and Habits.* University Press of Florida. Gainesville. 188 pp.

Bartlett, R. D., and P. P. Bartlett. 2011. *Florida's Turtles, Lizards, and Crocodilians: A Guide to Their Identification and Habits.* University Press of Florida. Gainesville. 257 pp.

Baxter, G. T., and M. D. Stone. 1980. *Amphibians and Reptiles of Wyoming.* Wyoming Game and Fish Department Bulletin 16. Cheyenne. 137 pp.

Bettelheim, M. P., R. Bury, L. C. Patterson, and G. M. Lubke. 2006. *Trachemys scripta elegans* (red-eared slider)—Reproduction in northern California. *Herpetological Review* 37: 459–460.

Boersma, P. D., S. H. Reichard, and A. N. Van Buren (eds.). 2006. *Invasive Species in the Pacific Northwest.* University of Washington Press, Seattle. 276 pp.

Boundy, J., and J. L. Carr. 2017. *Amphibians & Reptiles of Louisiana: An Identification and Reference Guide.* Louisiana State University Press, Baton Rouge. 386 pp.

Bradley, W. G., and J. E. Deacon. 1966. Amphibians and reptile records for southern Nevada. *Southwestern Naturalist* 11: 132–134.

Brennan, T. C., and A. T. Holycross. 2006. *A Field Guide to Amphibians and Reptiles in Arizona.* Arizona Game and Fish Department, Phoenix. 150 pp.

Bury, R. B. 1995a. Slider: *Trachemys scripta* (Schoeff). *In* R. M. Storm and W. P. Leonard (eds.), *Reptiles of Washington and Oregon*, p. 39. Seattle Audubon Society. 176 pp.

Bury, R. B. 1995b. Notes on turtles. *In* R. M. Storm and W. P. Leonard (eds.), *Reptiles of Washington and Oregon*, pp. 10–13. Seattle Audubon Society. 176 pp.

Bury, R. B. 2008. Do urban areas favor introduced turtles in western North America? *In* J. C. Mitchell, R. E. Brown, and B. Bartholomew (eds.), *Urban Herpetology*, pp. 343–345. Society for Study of Amphibians and Reptiles, Salt Lake City, UT.

Bury, R. B., and D. J. Germano. 2008. *Actinemys marmorata* (Baird and Girard, 1852): western pond turtle. *In* P. C. Pritchard and A.G.J. Rhodin (eds.), *Conservation Biology of Freshwater Turtles*, pp. 1.1–1.9. Chelonian Research Monographs 3.

Bury, R. B., and R. A. Luckenbach. 1976. Introduced amphibians and reptiles in California. *Biological Conservation* 1976: 1–14.

Bury, R. B., H. H. Welsh Jr., D. J. Germano, and D. Ashton (eds.). 2012. *Western Pond Turtle: Biology, Sampling Techniques, Inventory and Monitoring, Conservation and Management.* Northwest Fauna 7. 128 pp.

Cadi, A., and P. Joly. 2003. Competition for basking places between the endangered European pond turtle (*Emys orbicularis galloitalica*) and the introduced red-eared slider (*Trachemys scripta elegans*). *Canadian Journal of Zoology* 81: 1392–1398.

Cadi, A., V. Delmas, A. Prevot-Juillard, P. Joly, C. Pieau, and M. Girondot. 2004. Successful reproduction of the introduced slider turtle (*Trachemys scripta*) in the south of France. *Aquatic Conservation: Marine and Freshwater Ecosystems* 14: 237–246.

California Department of Fish and Wildlife https://nrm.dfg.ca.gov/FileHandler.ashx?DocumentID=86535&inline

Carl, G. C. 1968. *The Reptiles of British Columbia.* 3rd ed. British Columbia Provincial Museum Handbook 3. Victoria. 65 pp.

Collins, J. T., S. L. Collins, and T. W. Taggart. 2010. *Amphibians, Reptiles, and Turtles in Kansas.* Eagle Mountain Publishing, Eagle Mountain, Utah. 312 pp.

Conant, R., and J. T. Collins. 1991. *Peterson Field Guide to Reptiles and Amphibians of Eastern and Central North America.* 3rd ed. Houghton Mifflin, New York. 450 pp.

Cook, F. R., R. W. Campbell, and G. R. Ryder. 2005. Origin and current status of the Pacific Pond Turtle (*Actinemys marmorata*) in British Columbia. *Wildlife Afield* 2: 58–63.

Crother, B. I. (ed.). 2017. *Scientific and Standard English Names of Amphibians and Reptiles of North America North of Mexico, with Comments Regarding Confidence in Our Understanding*, pp. 1–102. SSAR Herpetological Circular 43.

Cunningham, H. R., and N. H. Nazdrowicz. 2018. *The Maryland Amphibian and Reptile Atlas.* John Hopkins University Press, Baltimore. 283 pp.

Degenhardt, W. G., C. W. Painter, and A. H. Price. 1996. *Amphibians and Reptiles of New Mexico.* University of New Mexico Press, Albuquerque. 431 pp.

Department of Agriculture and Food. 2009. Red-eared slider animal pest alert no. 6/2009. Department of Agriculture and Food, Western Australia.

Dixon, J. R. 2013. *Amphibians & Reptiles of Texas.* 3rd ed. Texas A&M University Press, College Station. 447 pp.

Dodd, C. K., Jr. 2013. *Frogs of the United States and Canada: Volume 1.* Johns Hopkins University Press, Baltimore. 460 pp.

Dodd, C. K., Jr. 2013. *Frogs of the United States and Canada: Volume 2.* Johns Hopkins University Press, Baltimore. 982 pp.

Ernst, C. H., and J. E. Lovich. 2009. *Turtles of the United States and Canada.* 2nd ed. John Hopkins University Press, Baltimore. 827 pp.

Fogell, D. D. 2010. *A Field Guide to the Amphibians and Reptiles of Nebraska.* University of Nebraska-Lincoln, Institute of Agriculture and Natural Resources. 158 pp.

Gibbs, J. P., A. R. Breisch, P. K. Ducey, G. Johnson, and J. Behler. 2007. *The Amphibians and Reptiles of New York State.* Oxford University Press, New York. 422 pp.

Graeter, G. J., K. A. Buhlmann, L. R. Wilkinson, and J. W. Gibbons (eds.). 2013. *Inventory and Monitoring: Recommended Techniques for Reptiles and Amphibians.* Partners in Amphibian and Reptile Conservation Technical Publication IM-1. Birmingham, AL. 321 pp.

Gregory, P. T., and R. W. Campbell. 1984. *The Reptiles of British Columbia*. Provincial Museum Handbook 44, Victoria, BC. 103 pp.

Guyer, C., M. A. Bailey, and R. H. Mount. 2018. *Lizards and Snakes of Alabama*. University of Alabama Press, Tuscaloosa. 397 pp.

Harris, H. S., Jr. 1975. Distributional survey (Amphibia/Reptilia): Maryland and the District of Columbia. *Bulletin of the Maryland Herpetological Society* 11: 73–167.

Hidalgo-Vila, J., C. Díaz-Paniagua, A. Ribas, M. Florencio, N. Pérez-Santigosa, and J. C. Casanova. 2009. Helminth communities of the exotic introduced turtle, *Trachemys scripta elegans* in southwestern Spain: Transmission from native turtles. *Research in Veterinary Science* 86: 463–465.

Holland, D. H. 1994. The Western Pond Turtle: Habitat and History. Unpublished final report DOE/BP-62137-1. Bonneville Power Administration, U.S. Department of Energy, and Wildlife Diversity Program, Oregon Department of Fish and Wildlife. Portland. Available at: http://www.fwspubs.org/doi/suppl/10.3996/012016-JFWM-005/suppl_file/fwma-08-01-05_reference+s2.pdf?code=ufws-site. Accessed 29 January 2018.

Hulse, A. C., C. J. McCoy, and E. J. Censky. 2001. *Amphibians and Reptiles of Pennsylvania*. Cornell University Press, Ithaca, NY. 419 pp.

Jensen, E. L., P. Govindarajulu, J. Madsen, and M. Russello. 2014. Extirpation by introgression? Genetic evidence reveals hybridization between introduced *Chrysemys picta* and endangered western painted turtles (*C. p. bellii*) in British Columbia. *Herpetological Conservation and Biology* 9: 342–353.

Jenson, J. B., C. D. Camp, J. W. Gibbons, and M. J. Elliott. 2008. *Amphibians and Reptiles of Georgia*. University of Georgia Press, Athens. 575 pp.

Johnson, T. R. 2000. *The Amphibians and Reptiles of Missouri*. 2nd ed. Missouri Department of Conservation, Jefferson City. 400 pp.

Kraus, F. 2009. *Alien Reptiles and Amphibians*. Springer, New York. 563 pp.

Krysko, K. L., J. P. Burgess, M. R. Rochford, C. R. Gillette, D. Cueva, K. M. Enge, L. A. Somma, J. L. Stabile, D. C. Smith, J. A. Wasilewski, G. N. Kieckhefer III, M. C. Granatosky, and S. V. Nielsen. 2011. Verified non-indigenous amphibians and reptiles in Florida from 1863 through 2010: Outlining the invasion process and identifying invasion pathways and status. *Zootaxa* 3028: 1–64.

Krysko, K. L., K. M. Enge, and P. L. Moler (eds.). 2019. *Amphibians and Reptiles of Florida*. University Press of Florida, Gainesville. 728 pp.

Krysko, K. L., L. A. Somma, D. C. Smith, C. R. Gillette, D. Cueva, J. A. Wasilewski, K. M. Enge, S. A. Johnson, T. S. Campbell, J. R. Edwards, M. R. Rochford, R. Thompkins, J. L. Fobb, S. Mullin, C. Lechwoicz, D. Hazelton, and A. Warren. 2016. New verified non-indigenous amphibians and reptiles in Florida through 2015, with a summary of over 152 years of introductions. *IRCF Reptiles and Amphibians: Conservation and Natural History* 23: 110–143.

Lannoo, M. J. (ed.). 2005. *Amphibian Declines: The Conservation Status of the United States*. University of California Press, Berkeley. 1094 pp.

LeClere, J. B. 2013. *A Field Guide to the Amphibians and Reptiles of Iowa*. ECO Herpetological Publishing & Distribution. Rodeo, NM. 349 pp.

Lever, C. 2003. *Naturalized Reptiles and Amphibians of the World*. Oxford University Press, New York. 318 pp

Linsdale, J. M. 1940. Amphibians and Reptiles in Nevada. 1938. *Proceedings of the American Academy of Arts and Sciences* 73: 197–257.

Martof, B. S., W. M. Palmer, J. R. Bailey, and J. R. Harrison III. 1980. *Amphibians and Reptiles of the Carolinas and Virginia.* University of North Carolina Press, Chapel Hill. 264 pp.

Matsuda, B. M., D. M. Green, and P. T. Gregory. 2006. *Amphibians and Reptiles of British Columbia.* Royal British Columbia Museum, Victoria. 266 pp.

McKeown, S. 1996. *A Field Guide to Reptiles and Amphibians in the Hawaiian Islands.* Diamond Head Publishing, Los Osos, CA. 172 pp.

Meshaka, W. E., Jr. 2001. *The Cuban Treefrog in Florida: Life History of a Successful Colonizing Species.* University Press of Florida, Gainesville. 191 pp.

Meshaka, W. E., Jr. 2011. *A Runaway Train in the Making: The Exotic Amphibians, Reptiles, Turtles, and Crocodilians of Florida.* Herpetological Conservation and Biology 6: 1–101.

Meshaka, W. E., Jr., and K. J. Babbitt (eds.). 2005. *Amphibians and Reptiles: Status and Conservation in Florida.* Krieger Publishing, Melbourne, FL. 317 pp.

Meshaka, W. E., Jr., B. P. Butterfield, and J. B. Hauge. 2004. *The Exotic Amphibians and Reptiles of Florida.* Krieger Publishing, Melbourne, FL. 155 pp.

Minton, S. A., Jr. 2001. *Amphibians & Reptiles of Indiana.* 2nd ed. Indiana Academy of Science, Indianapolis. 404 pp.

Mitchell, J. C. 1994. *The Reptiles of Virginia.* Smithsonian Institution Press, Washington, DC. 352 pp.

Moriarty, J. J., and C. D. Hall. 2014. *Amphibians and Reptiles in Minnesota.* University of Minnesota Press. Minneapolis. 370 pp.

Mount, R. H. 1975. *The Reptiles & Amphibians of Alabama.* Auburn University Agricultural Experiment Station, Auburn, AL. 347 pp.

Nussbaum, R. A., R. M. Storm, and E. D. Brodie III. 1983. *Amphibians and Reptiles of the Pacific Northwest.* Caxton Press, Caldwell, ID. 336 pp.

Palmer, W. M., and A. L. Braswell. 1995. *Reptiles of North Carolina.* University of North Carolina Press, Chapel Hill. 412 pp.

Polo-Cavia, N., A. Gonzalo, P. López, and J. Martín. 2010. Predator recognition of native but not invasive turtle predators by naïve anuran tadpoles. *Animal Behaviour* 80: 461–466.

Polo-Cavia, N., P. López, and J. Martín. 2011. Feeding status and basking requirements of freshwater turtles in an invasion context. *Physiology and Behavior* 105: 1208–1213.

Powell, R., R. Conant, and J. T. Collins. 2016. *Peterson Field Guide to Reptiles and Amphibians of Eastern and Central North America.* Houghton Mifflin Harcourt, New York. 494 pp.

Pyke, G., A. White, P. Bishop, and B. Waldman. 2002. Habitat-use by the Green and Golden Bell Frog *Litoria aurea* in Australia and New Zealand. *Australian Zoologist* 32(1): 12–31.

Rosen, P. C., C. R. Schwalbe, D. A. Parizek Jr., P. A. Holm, and C. H. Lowe. 1995. Introduced aquatic vertebrates in the Chiricahua region: Effects on declining native ranid frogs. *In* L. H. DeBano, P. H. Folliott, A. Ortega-Rubio, G. J. Gottfried, R. H. Hamre, and C. B. Edminster (eds.), *Biodiversity and Management of the Madrean Archipelago: The Sky Islands of Southwestern United States and Northwestern Mexico,* pp. 251–261.

US Forest Service, Rocky Mountain Forest and Range Experiment Station, Fort Collins, CO. 669 pp.

Spinks, P. Q., G. B. Pauly, J. J. Crayon, and H. B. Shaffer. 2003. Survival of the western pond turtle (*Emys marmorata*) in an urban California environment. *Biological Conservation* 113: 257–267.

Stebbins, R. C. 2003. *A Field Guide to Western Reptiles and Amphibians*. 3rd ed. Houghton Mifflin, New York. 533 pp.

Stebbins, R. C., and S. M. McGinnis. 2012. *Field Guide to Amphibians and Reptiles of California*. University of California Press, Berkeley. 538 pp.

Stebbins, R. C., and S. M. McGinnis. 2018. *Peterson Field Guide to Western Reptiles & Amphibians*. 4th ed. Houghton Mifflin Harcourt, New York. 560 pp.

Thomson, R. C., P. Q. Spinks, and H. B. Shaffer. 2010. Distribution and abundance of invasive Red-Eared Sliders (*Trachemys scripta elegans*) in California's Sacramento River basin and possible impacts on native Western Pond Turtles (*Emys marmorata*). *Chelonian Conservation and Biology* 9: 297–302.

Trauth, S. E., H. W. Robison, and M. V. Plummer. 2004. *The Amphibians and Reptiles of Arkansas*. University of Arkansas Press, Fayetteville. 421 pp.

USGS (U.S. Geological Survey). *Invasive Species Program*. https://www.usgs.gov/science/mission-areas/ecosystems/invasive-species-program.

USGS (U.S. Geological Survey). 2018. *Nonindigenous Aquatic Species Database*. Gainesville, FL. Available at: http://nas.er.usgs.gov/. Accessed 14 January 2018.

Werner, J. K., B. A. Maxell, P. Hendricks, and D. L. Flath (eds.). 2004. *Amphibians and Reptiles of Montana*. Mountain Press Publishing, Missoula, MT. 262 pp.

White, J. F., Jr., and A. W. White. 2002. *Amphibians and Reptiles of Delmarva*. Tidewater Publishers, Centreville, MD. 248 pp.

Wildlife Act. 1996. [RSBC 1996] Chapter 488. Current to January 24, 2018. Victoria, BC, Canada. Available at: http://www.bclaws.ca/Recon/document/ID/freeside/00_96488_01. Accessed 29 January 2018.

Wildlife Act Designation and Exemption Regulation. 1990. BC Reg. 168/90, O.C. 758/90. Deposited May 18, 1990. Current to January 23, 2018. Victoria, BC, Canada. Available at: http://www.bclaws.ca/Recon/document/ID/freeside/13_168_90#section3.1. Accessed 29 January 2018.

Wildlife Act General Regulation. 1982. [includes amendments up to B.C. Reg. 127/2017, September 1, 2017]. BC Reg. 340/82 O.C. 1491/82. Filed July 30, 1982, effective August 1, 1982. Current to January 23, 2018. Victoria, BC, Canada. Available at: http://www.bclaws.ca/civix/document/id/lc/statreg/340_82. Accessed 30 January 2018.

Wilson, L. D., and L. Porras. 1983. *The Ecological Impact of Man on the South Florida Herpetofauna*. University of Kansas Museum of Natural History, Special Publication 9. 89 pp.

JOURNALS

The list below comprises the journals from which we derived records and natural history information on the species in this book.

Applied Herpetology (years 2003–2009; discontinued)

Bioinvasions Records

Biological Conservation
Bishop Museum Occasional Papers
Bulletin of the Chicago Herpetological Society
Catalogue of American Amphibians and Reptiles
Collinsorum
Copeia
Dactylus
Georgia Journal of Science
Herpetological Conservation and Biology
Herpetological Review
Journal of the Arkansas Academy of Science
Journal of Herpetology
Journal of Kansas Herpetology
Journal of the Tennessee Academy of Science
Maryland Naturalist
Mississippi Academy of Sciences
Northeast Gulf Science
Northeastern Naturalist
Pacific Science
Reptiles and Amphibians: Conservation and Natural History
Southeastern Naturalist
Texas Journal of Science
Transactions of the Illinois State Academy of Science
Urban Naturalist
Western North American Naturalist (formerly *Great Basin Naturalist*)

About the Authors

R. Bruce Bury is emeritus scientist at the U.S. Geological Survey (USGS). Before joining USGS, he was the first full-time herpetologist hired by the U.S. Fish and Wildlife Service. Dr. Bury was stationed for five years at the U.S. National Museum, Washington, DC, and then for 18 years each in Colorado and Oregon. He is the author of more than 175 scientific publications plus five books or monographs on turtles and tortoises.

Suzanne L. Collins is a wildlife photographer of national stature whose striking color images have appeared in many magazines and books. Her photographs have been exhibited at the Natural History Museum at the University of Kansas, Prairie Park Nature Center (Wichita), Bronx Zoo Reptile House, Tennessee Aquarium (Chattanooga), and in the Smithsonian Institution Traveling Exhibit on Declining Amphibians. Her images now appear worldwide on numerous websites. She has also coauthored numerous books.

Malcolm L. McCallum is a research scientist in the Department of Agriculture and Natural Resources at Langston University. He is an American environmental scientist, conservationist, herpetologist, and natural historian and is known for his work on the impending sixth mass extinction, conservation culturomics, and herpetology. He is a founding governing board member of the herpetology journal *Herpetological Conservation and Biology*. His research has been covered by David Attenborough, *Discover* magazine, and other media outlcts.

Walter E. Meshaka Jr. is senior curator of the Section of Zoology and Botany at the State Museum of Pennsylvania, Harrisburg. Before that, he worked for the United States Department of the Interior. The primary focus of his research is the ecology of eastern North American and exotic herpetofauna. Published books include *The Cuban Treefrog in Florida: Life History of a*

Successful Colonizing Species, The Exotic Amphibians and Reptiles of Florida, and *Amphibians and Reptiles: Status and Conservation in Florida.* He is a cofounder of the *Journal of North American Herpetology.*

About the Contributors

Karen Beard is professor of conservation ecology at Utah State University. She conducts research on invasive species and climate change. She has been studying the coqui frog for over 20 years.

Brian Gratwicke is a conservation biologist and leads the amphibian conservation programs at the Smithsonian Conservation Biology Institute. Gratwicke's focus has been on building capacity to conserve amphibians in Appalachia and Panama, developing outreach and educational programs and exhibits to build public support for amphibian conservation, and research to develop tools that will allow scientists to reintroduce amphibians back into the wild.

Brent Matsuda is a wildlife biologist and herpetologist for Triton Environmental Consultants based in Vancouver, BC. He has worked on reptile and amphibian issues in Canada and the United States for more than 25 years and continues to address the growing invasiveness of exotic herpetofauna against the advantage of warming climatic conditions.

Frank Mazzotti is professor of wildlife ecology at the University of Florida. His research and extension programs focus on endangered and invasive species in South Florida and the Caribbean.

Michael R. Rochford of Santa Rosa, California, is a professional herpetologist who spent ten years researching invasive herpetofauna in the Greater Everglades. His focus included radio telemetry, survey, trapping, and diet studies of the Burmese Python and Argentine Giant Tegu as well as removal efforts of the Nile Monitor, Spectacled Caiman, and various chameleon species.

Jesse Rothacker is the president of Forgotten Friend Reptile Sanctuary of Manheim, Pennsylvania, a nonprofit rescue and educational organization. Its mission is to give reptiles a chance through fun educational outreach and the adoption of unwanted pet reptiles.

Index